AF696098

WITHDRAWN-UNL

Surface-Launched Acoustic Wave Sensors

CHEMICAL ANALYSIS

A SERIES OF MONOGRAPHS ON
ANALYTICAL CHEMISTRY AND ITS APPLICATIONS

Editor
J. D. WINEFORDNER

VOLUME 144

A WILEY-INTERSCIENCE PUBLICATION

JOHN WILEY & SONS, INC.

New York / Chichester / Weinheim / Brisbane / Singapore / Toronto

Surface-Launched Acoustic Wave Sensors

Chemical Sensing and Thin-Film Characterization

MICHAEL THOMPSON and DAVID C. STONE

Department of Chemistry
University of Toronto
Toronto, Canada

A WILEY-INTERSCIENCE PUBLICATION

JOHN WILEY & SONS, INC.

New York / Chichester / Weinheim / Brisbane / Singapore / Toronto

This text is printed on acid-free paper.

Copyright © 1997 by John Wiley & Sons, Inc.

All rights reserved. Published simultaneously in Canada.

Reproduction or translation of any part of this work beyond that permitted by Section 107 or 108 of the 1976 United States Copyright Act without the permission of the copyright owner is unlawful. Requests for permission or further information should be addressed to the Permissions Department, John Wiley & Sons, Inc., 605 Third Avenue, New York, NY 10158-0012.

Library of Congress Cataloging in Publication Data:

Thompson, Michael, 1942–
Surface-launched acoustic wave sensors : chemical sensing and thin-film characterization / Michael Thompson and David C. Stone.
p. cm. — (Chemical analysis ; v. 144)
"A Wiley-Interscience publication."
Includes bibliographical references
ISBN 0-471-12794-9 (cloth : alk. paper)
1. Chemical detectors. 2. Acoustic surface wave devices. 3. Thin film devices. I. Stone, David (David Charles) II. Title. III. Series.
TP159.C46T48 1997
660′.283—dc20 96-44729

Printed in the United States of America

10 9 8 7 6 5 4 3 2 1

To Our Wonderful Ladies, Heather and Tammy

CONTENTS

PREFACE

This book concerns the design, operation and application of devices capable of generating acoustic waves in the ultrasonic frequency range. The clear emphasis of the text is the study of chemical and/or biochemical systems imposed on the surface of such devices, whether operated in the gas- or liquid-phase, *i.e.* on acoustic wave chemical and biological sensors. The last ten years or so have seen a surge of interest in chemical and biological sensor technology, in which an analyte is trapped by a recognitive element at the sensor surface. Acoustic wave devices, including those based on surface-launched waves, have occupied a prominent role in this area. While some would argue that the acoustic, electrochemical, optical and semiconductor formats employed for such devices are competing technologies, the reality is that each type of sensor can be viewed as providing complementary information.

Unlike the other technologies mentioned above, acoustic wave devices have, until recently, primarily been developed as sensors for specific chemical species. This is in contrast to, say, optical techniques, which have been employed for the study of interfaces for some time. Indeed, the fiber-optic chemical sensor (or "optrode") is a comparatively recent innovation. Recently, however, there have been numerous reports of acoustic energy being employed to study thin film and interfacial properties. This area of application, not surprisingly, figures prominently in this book.

In terms of the literature reviewed within this text, we have made no attempt to include every single published paper and conference proceeding touching on this field. The informed reader will, for example, find little discussion of the so-called quartz crystal microbalance, especially its electrochemical applications. The reason for this is simply the huge volume of literature on this subject, which is deserving of a text in its own right. In fact, we have deliberately focussed on devices that employ acoustic waves *launched and received on the same surface.* We have, however, summarized the technical background behind this and a number of other acoustic wave sensors. The reader will also notice a degree of overlap between the reference lists at the end of each Chapter. This is again a deliberate action on our part, as we feel that this is a considerably more manageable format than having a single list at the end of the text. We have also been quite selective in terms of applications discussed. The interested reader will, however, find extensive references in the summary tables in Chapters 3 and 5.

Although disclaimers are often met with skepticism, or outright disdain, the authors wish to point out that this book has been written primarily with the larger community of chemists in mind. One of the hazards (and benefits) of chemical sensor research is that it blithely transcends many of the traditional disciplinary (and intra-disciplinary) boundaries of formal education and research. Thus we touch upon such seemingly diverse areas as acoustic wave physics, applied mathematics, chemistry, electronics, fluid mechanics, materials science, piezoelectricity and polymer science, each of which has its own unique language and viewpoints. We have, therefore, attempted to present all of this in a manner which is both self-consistent and readable for the non-expert, without sacrificing accuracy or trivializing the subject matter. We can only hope that the engineering community will forgive us for any lapses in this respect, and for essentially ignoring the vast areas of non-chemical application of acoustic wave devices. At the same time, we trust that the non-chemist will find the chemical sections of the text both informative and enlightening.

Finally, the authors wish to express their sincere gratitude for the loving support of their families through the sometimes trying experience of preparing this book. It would not have been possible without them. Tremendous support was also provided by the members of the research group of Michael Thompson and David Stone, who demonstrated considerable patience when the authors had to spend yet more time "on the book." Lastly, but by no means least, the authors are extremely grateful to Blanca Granozio, who spent many hours assisting us in the preparation of the manuscript. Her unflagging sense of humour and willingness to help were always a source of inspiration.

MICHAEL THOMPSON
DAVID C. STONE

Toronto, Canada

CHEMICAL ANALYSIS

A SERIES OF MONOGRAPHS ON ANALYTICAL CHEMISTRY AND ITS APPLICATIONS

J. D. Winefordner, *Series Editor*

Vol. 1. **The Analytical Chemistry of Industrial Poisons, Hazards, and Solvents.** *Second Edition.* By the late Morris B. Jacobs

Vol. 2. **Chromatographic Adsorption Analysis.** By Harold H. Strain (*out of print*)

Vol. 3. **Photometric Determination of Traces of Metals.** *Fourth Edition*
Part 1: General Aspects. By E. B. Sandell and Hiroshi Onishi
Part IIA: Individual Metals, Aluminum to Lithium. By Hiroshi Onishi
Part IIB: Individual Metals, Magnesium to Zirconium. By Hiroshi Onishi

Vol. 4. **Organic Reagents Used in Gravimetric and Volumetric Analysis.** By John F. Flagg (*out of print*)

Vol. 5. **Aquametry: A Treatise on Methods for the Determination of Water.** *Second Edition (in three parts).* By John Mitchell, Jr. and Donald Milton Smith

Vol. 6. **Analysis of Insecticides and Acaricides.** By Francis A. Gunther and Roger C. Blinn (*out of print*)

Vol. 7. **Chemical Analysis of Industrial Solvents.** By the late Morris B. Jacobs and Leopold Schetlan

Vol. 8. **Colorimetric Determination of Nonmetals.** *Second Edition.* By the late David F. Boltz and James A. Howell

Vol. 9. **Analytical Chemistry of Titanium Metals and Compounds.** By Maurice Codell

Vol. 10. **The Chemical Analysis of Air Pollutants.** By the late Morris B. Jacobs

Vol. 11. **X-Ray Spectrochemical Analysis.** *Second Edition.* By L. S. Birks

Vol. 12. **Systematic Analysis of Surface-Active Agents.** *Second Edition.* By Milton J. Rosen and Henry A. Goldsmith

Vol. 13. **Alternating Current Polarography and Tensammetry.** By B. Breyer and H. H. Bauer

Vol. 14. **Flame Photometry.** By R. Herrmann and J. Alkemade

Vol. 15. **The Titration of Organic Compounds** (*in two parts*). By M. R. F. Ashworth

Vol. 16. **Complexation in Analytical Chemistry: A Guide for the Critical Selection of Analytical Methods Based on Complexation Reactions.** By the late Anders Ringbom

Vol. 17. **Electron Probe Microanalysis.** *Second Edition.* By L. S. Birks

Vol. 18. **Organic Complexing Reagents: Structure, Behavior, and Application to Inorganic Analysis.** By D. D. Perrin

Vol. 19. **Thermal Analysis.** *Third Edition.* By Wesley Wm. Wendlandt

Vol. 20. **Amperometric Titrations.** By John T. Stock

Vol. 21. **Reflectance Spectroscopy.** By Wesley Wm. Wendlandt and Harry G. Hecht

Vol. 22. **The Analytical Toxicology of Industrial Inorganic Poisons.** By the late Morris B. Jacobs

Vol. 23. **The Formation and Properties of Precipitates.** By Alan G. Walton

Vol. 24. **Kinetics in Analytical Chemistry.** By Harry B. Mark, Jr. and Garry A. Rechnitz

Vol. 25. **Atomic Absorption Spectroscopy.** *Second Edition.* By Morris Slavin

Vol. 26. **Characterization of Organometallic Compounds** (*in two parts*). Edited by Minoru Tsutsui

Vol. 27. **Rock and Mineral Analysis.** *Second Edition.* By Wesley M. Johnson and John A. Maxwell

Vol. 28. **The Analytical Chemistry of Nitrogen and Its Compounds** (*in two parts*). Edited by C. A. Streuli and Philip R. Averell

Vol. 29. **The Analytical Chemistry of Sulfur and Its Compounds** (*in three parts*). By J. H. Karchmer

Vol. 30. **Ultramicro Elemental Analysis.** By Günther Tölg

Vol. 31. **Photometric Organic Analysis** (*in two parts*). By Eugene Sawicki

Vol. 32. **Determination of Organic Compounds: Methods and Procedures.** By Frederick T. Weiss

Vol. 33. **Masking and Demasking of Chemical Reactions.** By D. D. Perrin

Vol. 34. **Neutron Activation Analysis.** By D. De Soete, R. Gijbels, and J. Hoste

Vol. 35. **Laser Raman Spectroscopy.** By Marvin C. Tobin

Vol. 36. **Emission Spectrochemical Analysis.** By Morris Slavin

Vol. 37. **Analytical Chemistry of Phosphorus Compounds.** Edited by M. Halmann

Vol. 38. **Luminescence Spectrometry in Analytical Chemistry.** By J. D. Winefordner, S. G. Schulman and T. C. O'Haver

Vol. 39. **Activation Analysis with Neutron Generators.** By Sam S. Nargolwalla and Edwin P. Przybylowicz

Vol. 40. **Determination of Gaseous Elements in Metals.** Edited by Lynn L. Lewis, Laben M. Melnick, and Ben D. Holt

Vol. 41. **Analysis of Silicones.** Edited by A. Lee Smith

Vol. 42. **Foundations of Ultracentrifugal Analysis.** By H. Fujita

Vol. 43. **Chemical Infrared Fourier Transform Spectroscopy.** By Peter R. Griffiths

Vol. 44. **Microscale Manipulations in Chemistry.** By T. S. Ma and V. Horak

Vol. 45. **Thermometric Titrations.** By J. Barthel

Vol. 46. **Trace Analysis: Spectroscopic Methods for Elements.** Edited by J. D. Winefordner

Vol. 47. **Contamination Control in Trace Element Analysis.** By Morris Zief and James W. Mitchell

Vol. 48. **Analytical Applications of NMR.** By D. E. Leyden and R. H. Cox

Vol. 49. **Measurement of Dissolved Oxygen.** By Michael L. Hitchman

Vol. 50. **Analytical Laser Spectroscopy.** Edited by Nicolo Omenetto

Vol. 51. **Trace Element Analysis of Geological Materials.** By Roger D. Reeves and Robert R. Brooks

Vol. 52. **Chemical Analysis by Microwave Rotational Spectroscopy.** By Ravi Varma and Lawrence W. Hrubesh

Vol. 53. **Information Theory As Applied to Chemical Analysis.** By Karel Eckschlager and Vladimir Štěpánek

Vol. 54. **Applied Infrared Spectroscopy: Fundamentals, Techniques, and Analytical Problem-solving.** By A. Lee Smith

Vol. 55. **Archaeological Chemistry.** By Zvi Goffer

Vol. 56. **Immobilized Enzymes in Analytical and Clinical Chemistry.** By P. W. Carr and L. D. Bowers

Vol. 57. **Photoacoustics and Photoacoustic Spectroscopy.** By Allan Rosencwaig

Vol. 58. **Analysis of Pesticide Residues.** Edited by H. Anson Moye

Vol. 59. **Affinity Chromatography.** By William H. Scouten

Vol. 60. **Quality Control in Analytical Chemistry.** *Second Edition.* By G. Kateman and L. Buydens

Vol. 61. **Direct Characterization of Fine Particles.** By Brian H. Kaye

Vol. 62. **Flow Injection Analysis.** By J. Ruzicka and E. H. Hansen

Vol. 63. **Applied Electron Spectroscopy for Chemical Analysis.** Edited by Hassan Windawi and Floyd Ho

Vol. 64. **Analytical Aspects of Environmental Chemistry.** Edited by David F. S. Natusch and Philip K. Hopke

Vol. 65. **The Interpretation of Analytical Chemical Data by the Use of Cluster Analysis.** By D. Luc Massart and Leonard Kaufman

Vol. 66. **Solid Phase Biochemistry: Analytical and Synthetic Aspects.** Edited by William H. Scouten

Vol. 67. **An Introduction to Photoelectron Spectroscopy.** By Pradip K. Ghosh

Vol. 68. **Room Temperature Phosphorimetry for Chemical Analysis.** By Tuan Vo-Dinh

Vol. 69. **Potentiometry and Potentiometric Titrations.** By E. P. Serjeant

Vol. 70. **Design and Application of Process Analyzer Systems.** By Paul E. Mix

Vol. 71. **Analysis of Organic and Biological Surfaces.** Edited by Patrick Echlin

Vol. 72. **Small Bore Liquid Chromatography Columns: Their Properties and Uses.** Edited by Raymond P. W. Scott

Vol. 73. **Modern Methods of Particle Size Analysis.** Edited by Howard G. Barth

Vol. 74. **Auger Electron Spectroscopy.** By Michael Thompson, M. D. Baker, Alec Christie, and J. F. Tyson

Vol. 75. **Spot Test Analysis: Clinical, Environmental, Forensic and Geochemical Applications.** By Ervin Jungreis

Vol. 76. **Receptor Modeling in Environmental Chemistry.** By Philip K. Hopke

Vol. 77. **Molecular Luminescence Spectroscopy: Methods and Applications** (*in three parts*). Edited by Stephen G. Schulman

Vol. 78. **Inorganic Chromatographic Analysis.** By John C. MacDonald

Vol. 79. **Analytical Solution Calorimetry.** Edited by J. K. Grime

Vol. 80. **Selected Methods of Trace Metal Analysis: Biological and Environmental Samples.** By Jon C. VanLoon

Vol. 81. **The Analysis of Extraterrestrial Materials.** By Isidore Adler

Vol. 82. **Chemometrics.** By Muhammad A. Sharaf, Deborah L. Illman, and Bruce R. Kowalski

Vol. 83. **Fourier Transform Infrared Spectrometry.** By Peter R. Griffiths and James A. de Haseth

Vol. 84. **Trace Analysis: Spectroscopic Methods for Molecules.** Edited by Gary Christian and James B. Callis

Vol. 85. **Ultratrace Analysis of Pharmaceuticals and Other Compounds of Interest.** By S. Ahuja

Vol. 86. **Secondary Ion Mass Spectrometry: Basic Concepts, Instrumental Aspects, Applications and Trends.** By A. Benninghoven, F. G. Rüdenauer, and H. W. Werner

Vol. 87. **Analytical Applications of Lasers.** Edited by Edward H. Piepmeier

Vol. 88. **Applied Geochemical Analysis.** By C. O. Ingamells and F. F. Pitard

Vol. 89. **Detectors for Liquid Chromatography.** Edited by Edward S. Yeung

Vol. 90. **Inductively Coupled Plasma Emission Spectroscopy: Part I: Methodology, Instrumentation, and Performance; Part II: Applications and Fundamentals.** Edited by J. M. Boumans

Vol. 91. **Applications of New Mass Spectrometry Techniques in Pesticide Chemistry.** Edited by Joseph Rosen

Vol. 92. **X-Ray Absorption: Principles, Applications, Techniques of EXAFS, SEXAFS, and XANES.** Edited by D. C. Konnigsberger

Vol. 93. **Quantitative Structure–Chromatographic Retention Relationships.** Edited by Roman Kaliszan

Vol. 94. **Laser Remote Chemical Analysis.** Edited by Raymond M. Measures

Vol. 95. **Inorganic Mass Spectrometry.** Edited by F. Adams, R. Gijbels, and R. Van Grieken

Vol. 96. **Kinetic Aspects of Analytical Chemistry.** By Horacio A. Mottola

Vol. 97. **Two-Dimensional NMR Spectroscopy.** By Jan Schraml and Jon M. Bellama

Vol. 98. **High Performance Liquid Chromatography.** Edited by Phyllis R. Brown and Richard A. Hartwick

Vol. 99. **X-Ray Fluorescence Spectrometry.** By Ron Jenkins

Vol. 100. **Analytical Aspects of Drug Testing.** Edited by Dale G. Deutsch

Vol. 101. **Chemical Analysis of Polycyclic Aromatic Compounds.** Edited by Tuan Vo-Dinh

Vol. 102. **Quadrupole Storage Mass Spectrometry.** By Raymond E. March and Richard J. Hughes

Vol. 103. **Determination of Molecular Weight.** Edited by Anthony R. Cooper

Vol. 104. **Selectivity and Detectability Optimizations in HPLC.** By Satinder Ahuja

Vol. 105. **Laser Microanalysis.** By Lieselotte Moenke-Blankenburg

Vol. 106. **Clinical Chemistry.** Edited by E. Howard Taylor

Vol. 107. **Multielement Detection Systems for Spectrochemical Analysis.** By Kenneth W. Busch and Marianna A. Busch

Vol. 108. **Planar Chromatography in the Life Sciences.** Edited by Joseph C. Touchstone

Vol. 109. **Fluorometric Analysis in Biomedical Chemistry: Trends and Techniques Including HPLC Applications.** By Norio Ichinose, George Schwedt, Frank Michael Schnepel, and Kyoko Adochi

Vol. 110. **An Introduction to Laboratory Automation.** By Victor Cerdá and Guillermo Ramis

Vol. 111. **Gas Chromatography: Biochemical, Biomedical, and Clinical Applications.** Edited by Ray E. Clement

Vol. 112. **The Analytical Chemistry of Silicones.** Edited by A. Lee Smith

Vol. 113. **Modern Methods of Polymer Characterization.** Edited by Howard G. Barth and Jimmy W. Mays

Vol. 114. **Analytical Raman Spectroscopy.** Edited by Jeannette Graselli and Bernard J. Bulkin

Vol. 115. **Trace and Ultratrace Analysis by HPLC.** By Satinder Ahuja

Vol. 116. **Radiochemistry and Nuclear Methods of Analysis.** By William D. Ehmann and Diane E. Vance

Vol. 117. **Applications of Fluorescence in Immunoassays.** By Ilkka Hemmila

Vol. 118. **Principles and Practice of Spectroscopic Calibration.** By Howard Mark

Vol. 119. **Activation Spectrometry in Chemical Analysis.** By S. J. Parry

Vol. 120. **Remote Sensing by Fourier Transform Spectrometry.** By Reinhard Beer

Vol. 121. **Detectors for Capillary Chromatography.** Edited by Herbert H. Hill and Dennis McMinn

Vol. 122. **Photochemical Vapor Deposition.** By J. G. Eden

Vol. 123. **Statistical Methods in Analytical Chemistry.** By Peter C. Meier and Richard Zund

Vol. 124. **Laser Ionization Mass Analysis.** Edited by Akos Vertes, Renaat Gijbels, and Fred Adams

Vol. 125. **Physics and Chemistry of Solid State Sensor Devices.** By Andreas Mandelis and Constantinos Christofides

Vol. 126. **Electroanalytical Stripping Methods.** By Khjena Z. Brainina and E. Neyman

Vol. 127. **Air Monitoring by Spectroscopic Techniques.** Edited by Markus W. Sigrist

Vol. 128. **Information Theory in Analytical Chemistry.** By Karel Eckschlager and Klaus Danzer

Vol. 129. **Flame Chemiluminescence Analysis by Molecular Emission Cavity Detection.** Edited by David Stiles, Anthony Calokerinos, and Alan Townshend

Vol. 130. **Hydride Generation Atomic Absorption Spectrometry.** Edited by Jiri Dedina and Dimiter L. Tsalev

Vol. 131. **Selective Detectors: Environmental, Industrial, and Biomedical Applications.** Edited by Robert E. Sievers

Vol. 132. **High Speed Countercurrent Chromatography.** Edited by Yoichiro Ito and Walter D. Conway

Vol. 133. **Particle-induced X-Ray Emission Spectrometry.** By Sven A. E. Johansson, John L. Campbell, and Klas G. Malmqvist

Vol. 134. **Photothermal Spectroscopy Methods for Chemical Analysis.** By Stephen Bialkowski

Vol. 135. **Element Speciation in Bioinorganic Chemistry.** Edited by Sergio Caroli

Vol. 136. **Laser-Enhanced Ionization and Spectrometry.** Edited by John C. Travis and Gregory C. Turk

Vol. 137. **Fluorescence Imaging Spectroscopy and Microscopy.** Edited by Xue Feng Wang and Brian Herman

Vol. 138. **Introduction of X-Ray Powder Diffractometry** by Ron Jenkins and Robert L. Snyder

Vol. 139. **Modern Techniques in Electroanalysis.** Edited by Petr Vanýsek

Vol. 140. **Total Reflection X-Ray Fluorescence Analysis.** By Reinhold Klockenkamper

Vol. 141. **Spot Test Analysis: Clinical, Environmental, Forensic, and Geochemical Applications, Second Edition.** By Ervin Jungreis

Vol. 142. **The Impact of Stereochemistry on Drug Development and Use.** Edited by Hassan Y. Aboul-Enein and Irving W. Wainer

Vol. 143. **Macrocyclic Compounds in Analytical Chemistry.** Edited by Yury A. Zolotov

Vol. 144 **Surface-Launched Acoustic Wave Sensors: Chemical Sensing and Thin-Film Characterization.** By Michael Thompson and David C. Stone

Surface-Launched Acoustic Wave Sensors

CHAPTER

1

INTRODUCTION

Within the science of analytical chemistry there are two emerging trends, both of which are fundamentally related to the increasingly demanding requirements for the detection and measurement of chemical species. The first is the evolution of the so-called "hyphenated" techniques such as gas chromatography–mass spectrometry, wherein sample pretreatment, separation of the sample components, and analytical measurement are combined into a single, continuous process. Modern techniques are constantly improving the selectivity, sensitivity, and limits of detection obtainable with such instrumental methods, but the underlying philosophy remains that of the classical wet chemical techniques. Simply put, the sample is brought to the laboratory, processed, and analyzed. The second emerging trend is toward analytical instrumentation that can be taken *out* of the laboratory *to* the sample, and which is simple to operate and maintain. The driving force behind this trend is a desire to make the analytical measurement more timely and relevant, and to provide immediate feedback based on the results. This is perhaps most apparent in the biomedical field, although it is also important in environmental analysis and process control. A logical extension of this rationale is to make localized measurements *within* the sample or sample environment. We can summarize this paradigm as "putting the instrument into the chemistry, not the chemistry into the instrument." It is within this context that the development of chemical and biological sensors has largely taken place.

A good working definition of a sensor is a device that allows the transduction of chemical and/or physical properties at an interface into usable information. A chemical sensor is thus a device that generates an electrical signal which is a function of chemical identity and/or concentration. It also follows that a sensor can equally well be used to characterize the properties of the interfacial region itself. Chemical sensors thus have the potential to be employed in both fundamental studies of interfaces and the detection and measurement of chemical species. Currently available transduction methods are usually classified as electrochemical, acoustic, optical, and thermal. These categories are, in turn, open to considerable division and subdivision. Sensors based on the propagation of acoustic waves, for example, are predominantly piezoelectric, but may also involve device structures that are not piezoelectric. In this case, acoustic energy is transferred to the sensor structure to cause the appropriate vibrations. Similarly, there are many possible piezoelectric device structures, acoustic wave propagation modes, and areas of application. In this book, therefore, we restrict discussion primarily to those piezoelectric devices employed for chemical sensing and thin-film characterization that employ surface-launched acoustic waves. We begin, however, by providing an historical overview of piezoelectricity, acoustic wave devices, and chemical sensors.

1.1. PIEZOELECTRICITY AND THE GENESIS OF ACOUSTIC WAVE DEVICES

Any discussion of the use of acoustic waves in sensor technology must surely begin with the discovery and provenance of the phenomenon known as *piezoelectricity*. Cady (1) defines piezoelectricity as the "*electric polarization produced by mechanical strain in crystals belonging to certain classes, the polarization being proportional to the strain and changing sign with it.*" In fact, this statement only deals with half of a reversible process, that is, the direct effect. It is equally true that a particular crystal can become mechanically strained when subject to electrical polarization, which is referred to as the converse piezoelectric effect. Accordingly, the whole process can be described as being truly reversible.

Historically it appears that there was considerable confusion as to whether electrical observations associated with certain crystals were related to pyroelectric or piezoelectric phenomena. This was caused by the apparent lack of distinction between "genuine" piezoelectricity and that related to deformation caused by a change of temperature (the pyroelectric effect). Certainly, it appears that piezoelectricity was observed long before the incredible work of the Curie brothers (Plate 1) in the 1880s. These two scientists, however, have been universally credited with the discovery of piezoelectricity (1880). In their classical experiments conducted in the Laboratory of Mineralogy, Sorbonne, plates of various crystals (including quartz) were subjected to applied pressure, the resulting electrical effects being recorded on an electrometer connected by means of tin foil electrodes (2). Although discovered by the Curie brothers, the term piezoelectricity was actually introduced by Hankel just a year after the first announcement of confirmation of the phenomenon (3,4).

The converse piezoelectric effect was predicted on thermodynamic grounds by Lippmann (5) in the same year that the phenomenon name was coined by Hankel. Astonishingly, in the same year (1881) the Curies were

Pierre Curie, 1905

Jacques Curie, 1926

Plate 1. (Upper) Pierre Curie (1859–1906), photographed in 1905. (Lower) Jacques Curie (1855–1941), photographed in 1926. Reprinted by kind permission of Dover Publications, Inc., New York, NY.

Plate 2. John William Strutt, Third Baron Rayleigh (1842–1919). Photographed in 1901. Reprinted by kind permission of Edward Arnold and Company, London, UK.

able to verify that the application of electrical energy can result in crystal mechanical deformation. Another giant in the scientific world, Lord Kelvin (6), was responsible for a theory of piezoelectricity based on thermodynamic concepts. Also, Voigt (7) developed a rigorous theory involving electric vectors and elastic tensors. Hence, in the span of a few years the discovery, provenance, and theoretical aspects of piezoelectricity had been worked out. In 1885, Lord Rayleigh (8) (Plate 2) was able to confirm the existence of elastic vibrations of the surface of solid materials. This prophetic work is really at the heart of much of this text, which deals with sensor structures based on the propagation of surface acoustic waves.

As is often the case with scientific discoveries, a considerable induction period occurred before the pace of work on piezoelectricity began to quicken, particularly with respect to emerging applications. One of the first such applications was developed by Langevin (9), who showed that piezoelectric quartz plates could be used to emit and receive acoustic waves propagating under water. This work led to that highly developed field we now call *ultrasonics,* which incorporates many applications, such as material characterization and medical imaging. Just after the first world war, Cady (10,11) produced the *piezoelectric resonator* out of studies of the electrical behavior of structures operating under conditions of mechanical resonance. This eventually led to applications in radio transmission and filter technology. Since these early days, numerous applications of the so-called volume or bulk piezoelectric resonator have appeared. Examples include the construction of precision clocks and watches; the measurement of pressure, velocity, and acceleration; monitoring of deposition under vacuum; and chemical and biosensor technology. Other types of bulk wave devices based on Love, Lamb, or Stonely waves became important in the 1950s in the development of dispersive delay lines for application in radar and sonar.

The marriage of piezoelectricity with microelectronics began in the 1960s. In this period White and Voltmer (12) showed that photolithographic techniques could be employed to deposit interdigital transducers on piezoelectric substrates to directly excite and couple to elastic surface waves. This provided significant advantages over the existing comb transducer technology, and has resulted in the development of many different transducer designs with application to signal processing. Primarily, the enormous advances that have occurred in surface acoustic wave technology over the past 30 years have been the result of the ability to operate at high frequencies, typically up to the giga-Hertz range, and the tremendous convenience of association with microelectronic technology. The first applications involved the use of surface acoustic wave-based filters for radar and sonar. Perhaps not surprisingly, the main users of this technology are the various branches of the military. Many other developments have occurred in the field of telecommunications, where the surface acoustic wave device has become a crucial element in terms of operation as a filter, demodulator, and oscillator. For an excellent appraisal of the use of these devices in signal processing, the reader might consult the comprehensive work of Feldman and Hénaff (13). Finally, the accessibility of the surface of the device, the relative ease of manipulation of quartz surface chemistry, and the concepts already introduced with respect to bulk-wave devices inevitably lead to the main thrust of this book, that is, application as a sensor responsive to gas- and liquid-phase analytes.

1.2. THE ADVENT OF CHEMICAL SENSOR TECHNOLOGY

The architecture of a typical chemical sensor consists of a site for selective recognition of atomic, ionic, or molecular species coupled to some form of transducer. The purpose of the former is to recognize and select an analyte present in either the gas or liquid phase, possibly from within a large array of interfering species. [For a discussion of the theoretical aspects of molecular recognition as applied to chemical sensors the reader might consult the article by Thompson and co-workers (14).] Hopefully, the very presence of the analyte on the device "surface" will convey qualitative analytical information to the scientist, whereas the number of sites occupied by such an entity will yield a quantitative result. The role of the transducer is to translate the presence of the selected analyte species into a detectable physical signal which can, in turn, be subject to readout. The structure of such a device might involve direct interaction of the analyte moiety with an actual component of the transducer itself, or composition of a special membrane or film for recognitive purposes that does not necessarily contribute to the transducing signal. Potential applications of chemical sensor technology span an enormous range, including monitoring the external environment and workplace from an occupational health point of view, detection of chemical warfare agents, analysis of gas fuels, and determination of ions in drinking water. A number of these possibilities are discussed in this book.

A particular form of chemical sensor, the biosensor, has become almost a separate research and application-associated endeavor in view of its special potential for the analysis of biochemical species. Although the general architecture of the biosensor is similar to that described above, the molecularly recognitive component is more specifically a biomolecule such as an antibody, enzyme, nucleic acid strand, or molecular receptor. Biosensor technology is not in the main purview of this text so only limited discussion of this area will follow; however, the interested reader should consult the excellent and readable book by Elizabeth Hall (15) or the series edited by Tony Turner (16).

The history of the development of chemical sensor technology is a fascinating topic, repleat with both brilliant successes and extravagant claims. It would be conceded by most that the first true chemical sensor was the glass electrode which is, of course, selective for hydronium ions. This device, which is based upon a potentiometric measurement associated with selective ion exchange at a liquid-to-hydrated glass matrix interface, was developed in the 1930s. Again, following a significant induction period, it was not until the 1960s that other ion-selective electrodes, such as the calcium sensor, appeared (17). To the present day, this technology is still an active area of research, particularly with respect to new forms of achieving ion selectivity and device engineering. Moreover, unlike some other areas of the sensor field, a number of devices are available commercially and in widespread use.

The 1960s also saw the introduction of the semiconductor detector for inflammable gases such as town and liquid propane gas (18). This type of sensor is fabricated from a metal oxide such as SnO_2, and the mechanism of response is based on the induction of a resistivity change by adsorption of molecules to the material surface. During the past 30 years and more an enormous effort, notably in Japan, has been made to improve the selectivity and stability of the device. Various dopants, such as noble metals, and a number of different semiconducting materials have been examined. Important applications of this type of sensor include humidity detection and vehicle emission control systems.

The union of microelectronics and chemical sensor science took a step forward in 1970 with the introduction of the ion-sensitive field-effect transistor (ISFET) by Bergveld (19). This device, which has been the subject of intense research activity over many years, functions on the principle of alteration of the source-to-drain current of an FET by processes occurring at the gate of the transistor, ion exchange being the primary example. Interestingly, there was considerable discussion in the early days as to whether it was necessary to operate the ISFET with a reference electrode. The lack of such a requirement was cited by Bergveld as a key advantage in terms of the original application, which was the measurement of local action potentials in biological systems in the absence of the "biological noise" that would be recorded by a conventional two-electrode system. The 1970s also witnessed the introduction of other sensors constructed with a strong microelectronics component. In 1975, Lundström and co-workers (20) fashioned the gate of a MOSFET from palladium metal and demonstrated that the device could be used to sense hydrogen gas selectively (Fig. 1.1). The mechanism of opera-

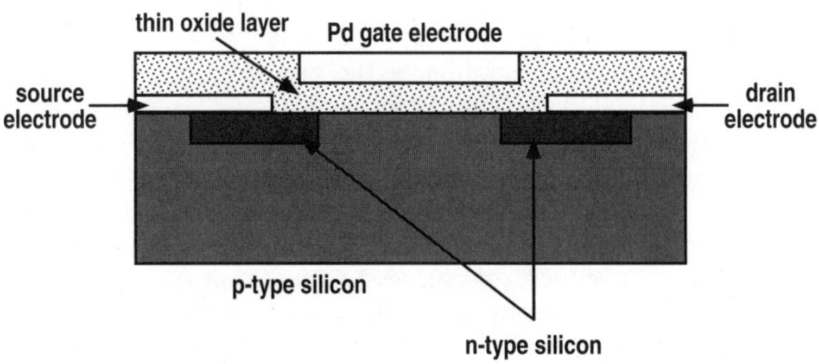

Figure 1.1. Cross section of a palladium-gate MOSFET hydrogen chemical sensor.

tion is based on the well-known solution of hydrogen in palladium, which ultimately results in a change in work function due to the adsorption of hydrogen atoms and, therefore, dipoles, at the metal–insulator interface. A new type of device called the charge-flow transistor (CFT) was introduced in 1977 for gas and humidity detection (21). Here, a resistive material is incorporated into a metal-free gap in the gate electrode of a MOSFET so that there is a time delay between application of the gate-to-source voltage and development of sufficient charge in the channel region to allow source-to-drain conduction. The time delay depends on film resistivity, and therefore on the presence of gas molecules at the film surface. This structure was the forerunner of the later monolithic CFT oscillator moisture sensor (22), an example of the successful integration of a chemical sensor with control and readout electronics in a single device.

The landmarks mentioned above are clearly dominated by semiconductor and/or electrochemical technology. Optical systems have been much more evident in recent years, however, particularly with respect to biosensors. Although the use of optical fibers was prevalent from the earliest development of such systems, the fibers were simply used to conduct light to and from conventional absorption or fluorescence experiments. Subsequently, efforts were made to immobilize an optically active reagent to the end of the fiber. The first such sensor was based on phenol red covalently bound to polymeric spheres, and was used to detect changes in pH (23). Later developments have involved immobilization of the appropriate chemistry on the fiber surface and the use of evanescent wave physics to probe the optical activity of bound reagents(s) (24). Fiber-optic sensors are claimed to posses a number of important advantages. These devices can allow the transmission of optical signals over considerable distances, which is important in remote sensing. They can also be isolated electronically at the chemical signalling end, which is highly important in the biomedical field.

A different type of optical sensor is based upon signals generated by surface plasmon resonance effects. In this case, the resonance angle is shifted when a sample is adsorbed to the surface of a (usually) silver or gold substrate. In 1982, an elegant demonstration of the principles of this method was published with respect to analysis of anesthetic gases (25). Absorption of the gas into a polymer film resulted in swelling which was detected by a surface-plasmon resonance sensor. Clearly, this experiment represents something of a harbinger of things to come, because it preceded the detection of biochemical interactions in the liquid phase at a sensor surface. This system forms the basis of a modern, commercially successful biosensor.

With respect to conferences and the review literature, the first international conference devoted solely to chemical sensors was held in Fukuoka, Japan in 1983 (26). The topic areas covered at the meeting were semiconductor, solid electrolyte, humidity, field-effect transistor, ion-selective biological, and other sensors. Notably, there was little in the way of sensors based on optical principles discussed at this conference. There have since been a number of these meetings, with the sixth held in the United States in 1996.

The first text to appear was that edited by Janata and Huber (27), which concentrated on solid-state devices. A subsequent volume edited by Edmonds (28) and published in 1988 contained a greater variety of sensor configurations (e.g., optical systems). A number of others, including conference proceedings, have appeared since the mid-1980s (29–35). Finally, new journals have appeared, together with sections of existing journals, which are devoted to sensor technologies. Examples of the former are *Sensors and Actuators,* chemistry version, and *Biosensors and Bioelectronics,* and the latter, *Talanta,* with its intermittent papers on chemical and biosensors.

This section has traced, albeit briefly, some of the earlier studies of chemical sensor science in order that the main subject matter of this text can be placed in context. Clearly, much has been omitted. One fact is evident from an appraisal of the history of the field, however, which is that practical and commercial success has been difficult to achieve. Significant sums have been spent on the development of liquid-phase FET devices, for example, yet today there is still no great commercial availability of this type of sensor. Despite considerable hyperbole we have not yet seen the claimed potential of the fiber-optic sensor being fully realized. Edmonds (28) was perhaps quite prophetic in his statement of 1988:

> The research chemist engaged in sensor development should be mindful of the fact that the ultimate target for these devices is the real world and that a successful laboratory device operating under well-defined conditions and careful calibration does not constitute a chemical sensor.

Possibly, this is as relevant today as it was in 1988.

1.3. ACOUSTIC WAVE DEVICES AND CHEMICAL SENSOR TECHNOLOGY

In 1895 Marie Sklodowska married her teacher, Pierre Curie, during her graduate work at the Sorbonne. In a scientific union, the Curies discovered the new element radium and in doing so were the first to use what they termed a "quartz electric balance." This was used to measure the charge stored in a capacitor as a result of the passage of ionizing radiation between its plates. Although a little unusual when considered in light of modern sensor practice, this experiment undoubtedly constitutes the first application of a piezoelectric material as a chemical sensor. A fascinating account of this work can be found in the thesis of Marie Curie (36), which was submitted to the Faculty of Sciences, Paris.

Direct deposition of films on the surface of a bulk acoustic wave (BAW) sensor had to wait until 1959 and the work of Sauerbrey (37). For AT-cut piezoelectric crystals, the resonant condition corresponds to a thickness shear oscillation in which the shear wave propagates through the bulk of the material, perpendicular to the faces of the crystal. The atomic displacements corresponding to this shear motion are thus parallel to the crystal surface. If material is deposited on either one or both faces of the crystal, the resonant frequency decreases. The first quantitative investigation of this effect was made by Sauerbrey, whose analysis was based on the increase in acoustic wavelength at resonance resulting from an increase in device thickness. For a rigidly adhered, thin film deposited in a vacuum, he derived the famous relationship between the frequency change, Δf (Hz), and the added mass, m_f (g):

$$\Delta f = -\frac{2f_0^2}{\sqrt{\mu_s \rho_s}} \frac{m_f}{A} \tag{1.1}$$

where f_0 is the fundamental resonant frequency of unloaded device (Hz), μ_s is the shear modulus of the piezoelectric substrate, ρ_s is the substrate density, and A is the surface area (cm^2). Note that m_f is the total mass added to both faces of the crystal, but A is the area of only one face. Collecting the constants together and letting $\Delta M = m_f/A$ gives

$$\Delta f = -C_1 \, \Delta M \tag{1.2}$$

which clearly predicts a linear frequency decrease with added mass per unit area, leading to the alternative name of quartz crystal microbalance (QCM). For AT-cut quartz, $\mu_s = 2.947 \times 10^{11}$ g cm^{-1} s^{-2} and $\rho_s = 2.648$ g cm^{-3}, and so $C_1 = 2.26 \times 10^{-6} f_0^2$ Hz cm^2 g^{-1} which, for typical operating frequencies, predicts a detection limit of the order of 10^{-12} g. Throughout the 1960s and 1970s a number of theoretical models of the mass deposition experiment on bulk acoustic wave devices were published (see review in reference 38). These included both consideration of the acoustic impedance of the added film and, in the case of liquid films, its viscosity.

The first report of a BAW device being used as a selective analytical sensor was published in 1964 by King (39). In this work, the device was coated with typical stationary-phase materials and incorporated into a gas–liquid chromatographic instrument as a detector. This theme has continued over many years and sensors have been produced for a number of gas-phase species such as NH_3, SO_2, CO_2, aromatic hydrocarbons, and pesticides. A significant contributor to this area has been Guilbaut and co-workers, who reviewed the field in 1984 (40). With respect to liquid-phase operation, prior to the late 1970s it was quite commonly believed that damping viscous forces would preclude employment of the device as a sensor in this phase. However, in 1980 both Nomura (41) and Bastiaans (42) illustrated that the device could successfully be operated in contact with a liquid if one face of the sensor was exposed to air. Since that time, a considerable body of work has been published with respect to both theoretical and practical aspects of liquid-phase operation. This was reviewed in 1991 by Thompson and co-workers (38). From an analytical viewpoint, the trend continues to this day that Sauerbrey's expression is invoked to explain frequency shifts on the addition or loss of material from the sensor surface when immersed in liquid. This is done despite the fact that, unlike the case for gas-phase operation, significant losses of acoustic energy occur by transmission to the surrounding medium.

Turning to the devices which are the subject of this text, the work of White and Voltmer referred to above initiated the new field of surface acoustic wave engineering. In its simplest form, oscillating electrical fields are induced between pairs of electrode fingers. These fields in turn create particle displacements in a piezoelectric material of correct crystal orientation. A surface acoustic wave (SAW) can be generated possessing longitudinal and shear components (a Rayleigh wave); this is detected by a second interdigital transducer. As with the bulk-wave device, the first description of a SAW chemical sensor—published by Wohltjen and Dessy in 1979—involved deposition of a sorptive film on the surface of the device (43,44). It was quickly established that the mass sensitivity of the SAW device is superior to the bulk device because it can be operated at significantly higher frequencies. The potential of the SAW device to characterize thin polymer films was also demonstrated (45). In the landmark paper by Wohltjen (46), it was argued that an absolute mass limit-of-detection of 3×10^{-15} g could be achieved. To this day, it would be fair to say that we still await experimental verification of this value. Again, as with the bulk-wave device, the following years saw attempts to enhance the selectivity of the device with respect to adsorption of gas-phase analyte molecules. In 1987, Nieuwenhuizen and Barendsz (47) carefully reviewed the compromise between the selectivity of the device and reversibility of adsorption processes. Additionally, the use of phthalocyanine films—which have figured prominently in chemical sensor research—as an adsorbing medium was discussed. In the same year, an *IEEE Transactions* issue was devoted entirely to acoustic wave sensor applications in which the surface wave variety was quite evident (48). Finally, with respect to sensor response, Thompson and co-workers (49) demonstrated for the first time that signals could be correlated with molecular orbital calculations of surface site-analyte molecule interactions.

In recent years, it has been recognized that properties of films other than their mass are also very important in determining the response of the device. Examples are the conductivity and viscoelasticity of the layers present on the sensor surface. Much of this work is discussed in later chapters of this book. With regard to liquid-phase operation, the surface Rayleigh wave device has had a much more checkered history than the bulk-wave sensor. The first paper to propose work in the liquid phase involved measurement of antibody–antigen reactions in situ at the device surface (50). A subsequent article, however, cast doubt as to whether the sensor was operating in a true Rayleigh wave mode (51).

Finally, the advances made over the years with respect to the two aforementioned sensors have certainly been partly responsible for initiating efforts to place other acoustic wave structures in chemical or biosensor format. In 1982, White and co-workers (52) introduced the thin-membrane surface-acoustic wave sensor. This device, which is based on various Lamb wave modes, has been used as a detector in both gas and liquid phases. Although the waves are surface-launched, the whole membrane is subject to particle displacement; accordingly, it is often termed an acoustic plate-mode (APM) device. Other plate-mode devices such as the SH (53) and Love wave (54,55) sensor have been employed for liquid-phase applications. Recent introductions have been the surface-transverse wave device (56) and sensors based on propagation of various modes of vibration in thin rods and tubes (57). In the former structure, Love waves are "trapped" using a dielectric film as waveguide. The acoustic waves propagated in thin rods and tubes are not a direct result of piezoelectric processes, but are generated by ultrasonic transducers coupled through glass or metal exponential horns. Despite the introduction of these new devices, the bulk acoustic and Rayleigh wave sensors still heavily dominate the chemical sensor literature, probably as a result of the limited commercial availability of the new devices.

1.4. BRIEF OVERVIEW OF THE TEXT

In this chapter we have attempted to provide a concise view of the landmarks in the development of piezoelectric devices as chemical sensors. Chapter 2 deals in more detail with the basic physics of piezoelectric materials and the nature of electromechanically induced vibrations in solids. This leads naturally to a phenomenological description of acoustic vibrations and their various modes in the different structures employed for sensor work. In Chapter 3, the effect of various factors such as temperature, pressure, potential, surface morphology, and added film conductivity and viscoelasticity are discussed from a theoretical basis. Included here is an evaluation of the effects of molecular sorption into thin films which is, of course, highly relevant to chemical sensor science. Chapter 4 deals with the basic theory of instigation of surface acoustic waves by interdigital transducers. This leads naturally into a discussion of device fabrication and appropriate measurement techniques, as well as practical issues relating to the electronic aspects of piezoelectric chemical sensors. This includes modes of operation, electronic circuitry, and measurement parameters. Chapter 5 goes on to discuss sensor response theory, which leads naturally into a discussion of the practical aspects of using acoustic wave devices in different sensor configurations. This includes temperature control, the influence of humidity, and the nature and deposition of the chemical coating required to achieve selective sensor response. Chapter 6 is highly application oriented, dealing with issues such as mass versus analytical sensitivity, single-sensor detection of particular analyte molecules, industrial and environmental applications, and a critical appraisal of sensor array techniques. With respect to the latter, we discuss the pros and cons of such applications as the detection of inorganic and organic gases and vapors, thin-film analysis, and biosensor development, and provide an appraisal of sensor-array techniques. The chapter concludes with a discussion of future directions, such as new approaches to coating design, devices, and the role of acoustic sensors in examining thin films and interfaces.

REFERENCES

1. W. G. Cady, *Piezoelectricity,* Vol. 1, Dover Publications, New York, 1964, p. 4.
2. J. Curie and P. Curie, *Bull. Soc. Min. France,* **1880,** *3,* 90.
3. W. G. Hankel, *Abh. Sächs,* **1881,** *12,* 547.
4. W. G. Hankel, *Ber. Sächs,* **1881,** *33,* 52.
5. G. Lippmann, *An. Chim. Phys. Ser. 5,* **1881,** *24,* 145.
6. Lord Kelvin, *Phil. Mag.,* **1893,** *36,* 331, 342, 384, 453.
7. W. Voight, *Abh. Götl,* **1890,** *36,* 1.
8. Lord Rayleigh, *Proc. London Math. Soc.,* **1885,** *17.*
9. A. Langevin and A. Moulin, *J. Phys.,* **1937,** *8,* 257.
10. W. G. Cady, *Phys. Rev.,* **1921,** 17, 531.
11. W. G. Cady, *Proc. Inst. Radio Eng.,* **1922,** *10,* 83.
12. R. M. White and F. W. Voltmer, *Appl. Pys. Lett.,* **1965,** *7,* 314.
13. M. Feldman and J. Hénaff, *Surface Acoustic Waves for Signal Processing,* Artech House, Boston, MA, 1989.

14. M. Thompson, M. D. Frank, W. M. Heckl, F. M. Marassi and S. J. Vigmond in *Chemical Sensor Technology,* Vol. 2, T. Seiyama, Ed., Kodansha/Elsevier, Tokyo, 1989, pp. 237–254.

15. E. A. Hall, *Biosensors,* Open University Press, Buckingham, UK, 1990.

16. A. P. F. Turner, Ed., *Advances in Biosensors,* Vols. 1–3, JAL Press, London, UK, 1991–1995.

17. J. W. Ross, *Science,* **1966,** *156,* 1378.

18. T. Seiyama, A. Kato, K. Fujiishi, and M. Nagatani, *Anal. Chem.,* **1966,** *38,* 1069.

19. P. Bergveld, *IEEE Trans. Biomed Eng.,* **1970,** *BME-19,* 70.

20. K. I. Lundström, M. S. Shivaraman, and C. N. Svensson, *J. Appl. Phys.,* **1975,** *46,* 3876.

21. S. D. Senturia, C. M. Sechen, and J. A. Wishneusky, *Appl. Phys. Lett.,* **1977,** *30,* 106.

22. S. D. Senturia, S. L. Garverick, and K. Togashi, *Sens. Actuators,* **1981/82,** *2,* 59.

23. J. I. Peterson, S. R. Goldstein, R. V. Fitzgerald, and D. K. Buckhold, *Anal. Chem.,* **1980,** *52,* 864.

24. R. M. Sutherland, C. Dahne, J. F. Place, and A. S. Ringrose, *J. Immunol. Methods,* **1984,** *74,* 253.

25. C. Nylander, B. Liedberg, and T. Lind, *Sens. Actuators,* **1982,** *3,* 79.

26. T. Seiyama, K. Fueki, J. Shiokawa, and S. Suzuki, *Proc. Intl. Meeting, Chem. Sensors,* held in Fukuoka, Japan, Sept. 19–22, 1983, Kodansha/Elsevier, Tokyo, 1983.

27. J. Janata and R.J. Huber, Eds., *Solid State Chemical Sensors,* Academic Press, New York, 1985.

28. T. E. Edmonds, Ed, *Chemical Sensors,* Blackie, London, 1988.

29. J. Janata, *Principles of Chemical Sensors,* Plenum Press, New York, 1989.

30. D. Schuetzle and R. Hammerle, Eds., *Fundamentals and Applications of Chemical Sensors,* American Chemical Society, Washington, DC, ACS Symposium Series, 1986, Vol. *309.*

31. R. W. Murray, R. E. Dessy, W. R. Heineman, J. Janata, and W. R. Seitz, Eds., *Chemical Sensors and Microinstrumentation,* ACS Symposium Series, American Chemical Society, Washington, DC, 1989, Vol. *403.*

32. T. Seiyama, Ed., *Chemical Sensor Technology,* Vols. 1 and 3, Kodansha/Elsevier, Tokyo, 1988 and 1989.

33. P. G. Edelman and J. Wang Eds., *Biosensors and Chemical Sensors,* American Chemical Society, ACS Symposium Series, Washington, DC, 1992, Vol. *487.*

34. T. E. Mallouk and J. Harrison Eds., *Interfacial Design and Chemical Sensing,* American Chemical Society, ACS Symposium Series, Washington, DC, 1994, Vol. *561.*

35. J. W. Gardner and P. N. Bartlett, Eds., *Sensors and Sensory Systems for an Electronic Nose,* NATO ASI Series, Series E: Applied Sciences, Vol. 212, Kluwer Publishing, London, 1992.

36. M. Curie, *Radioactive Substances* (a translation of the Thesis presented to the Faculty of Sciences, Paris), Philosophical Library Inc., New York, 1961.
37. G. Sauerbrey, *Z. Phys.*, **1959,** *155,* 206.
38. M. Thompson, A. L. Kipling, W. C. Duncan-Hewitt, Lj. V. Rajakovic, and B. A. Cavic-Vlasak, *Analyst,* **1991,** *116,* 881.
39. W. H. King, *Anal. Chem.*, **1964,** *36,* 1735.
40. G. G. Guilbaut, in *Applications of Piezoelectric Quartz Crystal Microbalances,* Vol. 7, C.-S. Lu and A.W. Czanderna, Eds., Elsevier, New York, **1984,** p. 251.
41. T. Nomura and A. Minemura, *Nippon Kagaku Kaishi,* **1980,** *1980,* 1621.
42. P. L. Konash and G. J. Bastiaans, *Anal. Chem.*, **1980,** *52,* 1929.
43. H. Wohltjen and R. Dessy, *Anal. Chem.*, **1979,** *51,* 1458.
44. H. Wohltjen and R. Dessy, *Anal. Chem.*, **1979,** *51,* 1465.
45. H. Wohltjen and R. Dessy, *Anal. Chem.*, **1979,** *51,* 1470.
46. H. Wohltjen, *Sens. Actuators,* **1984,** *5,* 307.
47. M.S. Nieuwenhuizen and A.W. Barendsz, *Sens. Actuators,* **1987,** *11,* 45.
48. *IEEE Trans. Ultrason., Ferroelec. Freq. Control,* **1987,** *UFFC-34(2).*
49. W. M Heckl, F. M. Marassi, K. M. R. Kallury, D. C. Stone, and M. Thompson, *Anal. Chem.*, **1990,** *62,* 32.
50. J. E. Roederer and G. J. Bastiaans, *Anal. Chem.*, **1983,** *55,* 2333.
51. G. S. Calabrese, H. Wohltjen and M. K. Roy, *Anal. Chem.*, **1987,** *59,* 833.
52. C. T. Chuang, R. M. White, and J. J. Bernstein, *IEEE Electron Device Lett.*, **1982,** *EDL-3,* 145.
53. S. J. Martin, A. J. Ricco, T. M. Niemczyk, and G. C. Frye, *Sens. Actuators,* **1989,** *20,* 253.
54. T. Nomura and T. Yasuda, *IEEE Ultrason. Symp. (Proc.),* **1990,** 307
55. C. Caliendo, A. D'Amico, P. Verardi, and W. Venema, *IEEE Ultrason. Symp. (Proc.),* **1990,** 383
56. R. L. Baer, C. A. Flory, M. Tom-Moy, and P. S. Solomon, *IEEE Ultrason. Symp. (Proc.),* **1992,** 293.
57. P. C. H. Li, D. C. Stone, and M. Thompson, *Anal. Chem.*, **1993,** *65,* 2177.

CHAPTER

2

ACOUSTIC WAVES IN SOLIDS AND DEVICE STRUCTURE

The launching of acoustic waves in solids at ultrasonic frequencies (of the order of one to several hundred mega-Hertz) depends on the conversion of electromagnetic into acoustic energy. This process may involve, for example, inherent electromechanical mechanisms in piezoelectric materials or coupling of nonpiezoelectric solids with transducers capable of generating acoustic fields. The physics of propagation of acoustic waves in various materials is exceedingly complex in that it involves crystallography, electromechanical phenomena, wave engineering, and tensor mathematics, among a number of subdisciplines. Since this text is largely concerned with the use of acoustic wave sensors in analytical chemistry, this chapter is deliberately concise in its somewhat phenomenological treatment of the necessary crystal physics that underpins acoustic device structure. The reader can access a large volume of textual literature on piezoelectricity and acoustic wave physics for much further detail (1–5). The emphasis in this section is on the design and format of devices that have been developed for application in analytical chemistry. An important element is an attempt to avoid confusion with respect to terminology and description of the nature of the acoustic waves propagating in these devices, which has perhaps plagued the field in the past. Since it is often the case that tensor mathematics can alienate a reader from the chemistry community, this is kept to a minimum.

2.1. ELASTICITY OF SOLIDS

The various sensors to be described in this text have in common that a solid component of the device, piezoelectric or otherwise, is deformed by the application of a stress. The latter is caused by an electric field or by acoustic energy emanating from a transducer. Accordingly, a discussion of the propagation of acoustic waves must begin with a description of particle displacement in a solid (deformation) caused by a stress. To develop the basic equations of elasticity, let us first briefly review the concept of a stress tensor.

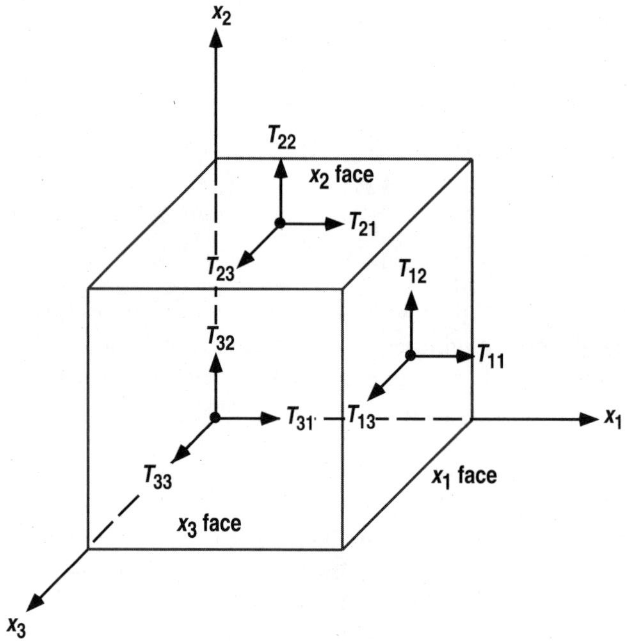

Figure 2.1. Stresses at points in space.

Consider a cube of solid material as shown in Figure 2.1. Note that the numbers 1, 2, and 3 label the x, y and z directions, respectively, and that the possible stresses are denoted as T_{ij}. The first subscript denotes the face on which the stress acts, and the second the direction of stress on the face. There is an array of nine stresses at a location in space, which is represented by the stress tensor

$$T_{ij} = \begin{bmatrix} T_{11} & T_{12} & T_{13} \\ T_{21} & T_{22} & T_{23} \\ T_{31} & T_{32} & T_{33} \end{bmatrix} \tag{2.1}$$

This tensor must be symmetric (due to the conservation of angular momentum at each point), that is, $T_{ij} = T_{ji}$ and, therefore, there are actually six independent components of the tensor. It is termed a second-rank tensor because it possesses two subscripts.

The theory of elasticity begins with the linearized form of Newton's second law for a viscoelastic solid of density ρ_m:*

$$\rho_m \ddot{u}_j = T_{ij,i} + F_{(m)j} \tag{2.2}$$

where u_j is the particle displacement and $F_{(m)j}$ is the external force. Because of the symmetry of the stress tensor mentioned above, Equation 2.2 contains nine dependent variables.

The application of a stress on a viscoelastic medium causes a strain which is represented by the strain tensor:

$$S_{ij} = \frac{1}{2}(u_{ij} + u_{ji}) \tag{2.3}$$

As was the case for the stress tensor, because of symmetry, Equation 2.3 contains nine dependent variables. Now we are in a position to express the link between stress and strain:

$$T_{ij} = c_{ijkl} S_{kl} + \eta_{ijkl} \dot{S}_{kl} \tag{2.4}$$

This famous equation describes, in its more general form, Hooke's law for a viscoelastic medium. The coefficients c_{ijkl} and η_{ijkl} are, respectively, the fourth-rank elastic coefficient tensor and the fourth-rank viscoelastic coefficient tensor. Because of the symmetry of the stress and strain tensors alluded to above, c_{ijkl} and η_{ijkl} also satisfy the following relations:

$$c_{ijkl} = c_{ijlk} = c_{jikl} = c_{klij} \tag{2.5}$$

$$\eta_{ijkl} = \eta_{ijlk} = \eta_{jikl} = \eta_{klij} \tag{2.6}$$

Furthermore, there exist 21 independent values of c_{ijkl}. A number of devices to be described presently are composed of the piezoelectric material quartz, and to give the reader a feeling for the magnitude of the elastic coefficients, values for this material are included Table 2.1 (3).

If we combine Newton's second law of motion with the equations for stress and strain such that the latter two parameters are eliminated, the result is a general viscoelastic wave equation:

$$c_{ijkl} u_{k,li} + \eta_{ijk} \dot{u}_{k,li} + F_{(m)j} = \rho_m \ddot{u}_j \tag{2.7}$$

* Tensor notation is used in this section of the text. The indices i, j, and k represent the x, y, and z directions and take the values 1, 2, and 3. Subscript i that occurs once in a term takes all three values. Presence of a comma in a subscript indicates the operation of partial differentiation with respect to a coordinate. A single dot above a symbol implies single differentiation, while a double dot implies double differentiation.

TABLE 2.1. Elastic Coefficients c_{ijkl} for Quartz

Component	Value ($\times$ 10^9 N m^{-2})	Component	Value ($\times$ 10^9 N m^{-2})
c_{1111}	86.74	c_{2222}	129.77
c_{1112}	0.0	c_{2223}	5.7
c_{1113}	0.0	c_{2233}	−7.42
c_{1122}	−8.25	c_{2312}	0.0
c_{1123}	−3.66	c_{2313}	0.0
c_{1133}	27.15	c_{2323}	38.61
c_{1212}	29.01	c_{3312}	0.0
c_{1312}	2.53	c_{3313}	0.0
c_{1313}	68.81	c_{3323}	9.92
c_{2212}	0.0	c_{3333}	102.83
c_{2213}	0.0		

This equation can be solved to generate the particle displacement u_i obtained for application of stress. This form of wave motion will be highly relevant to devices which are constructed from nonpiezoelectric materials.

2.2. PIEZOELECTRICITY

To develop wave equations for piezoelectric materials and, therefore, the various sensors constructed from these solids, we must couple the equations of elasticity with the equations of electromagnetism. Before setting out on this course, however, we must digress for a time and introduce the nature of piezoelectric materials so that the origins of the electromechanical processes can be traced.

2.2.1. Piezoelectric Materials

Piezoelectricity is associated with the instigation of electric polarization when a particular material is subjected to mechanical stress, and with the onset of deformation in the material when placed in an electric field. The origin of the property lies in the interaction of Coulombic charge forces and elastic restoring forces that arise under such circumstances.

An all-important factor in determining whether a particular material exhibits a piezoelectric effect is the crystallographic nature of the solid. Piezoelectricity cannot exist in solids possessing *central symmetry*. The last statement clearly implies that we are concerned with that three-dimensional array of atoms we call a crystal, rather than with amorphous solids. Accordingly, a brief review of basic crystallography, with its particular notation systems, is necessary to evaluate which crystals exhibit piezoelectricity. The *space lattice* of a crystal is composed of a large number of identical *unit cells.* The latter constitutes a fully representative component of the lattice.

Turning to symmetry arguments, we must now define the complete set of symmetry elements that exist for a space lattice (shorthand notation is included in parentheses). There are onefold (1), twofold (2), threefold (3), fourfold (4), and sixfold (6) rotational axes, center-of-symmetry operation ($\bar{1}$), rotation inversion ($\bar{3}$, $\bar{4}$, $\bar{6}$), and mirror-plane operations (m), counting 10 in all. Crystals can be categorized into seven *crystal systems,* depending on the level of symmetry exhibited. These systems can be subdivided into other classes—called *point groups*—of which there are 32 in all (Table 2.2). Each point group can be represented by a set of symbols. For example, 6/m denotes a sixfold rotational axis with a mirror plane at right angles to it. This notation

TABLE 2.2. Crystallographic Systems with Point Groups in Full Hermann–Maugin Notation

System	Essential Symmetry	Point Groups[a]	Examples
Cubic	4 threefold, 3 twofold, or 3 fourfold axes	(5), 23, 432, 2/m$\bar{3}$, $\bar{4}$3m, 4/m$\bar{3}$2/m,	CaF_2; ZnS, NaCl, diamond, Pb
Tetragonal	1 fourfold axis	(7), 4mm, 4/m2/m2/m, $\bar{4}$2m, 422, 4/m, 4, $\bar{4}$	SnO_2, TiO_2, $PbWO_4$
Orthorhombic	3 threefold axes or 1 twofold axis and 2 perpendicular planes in a twofold axis	(3), 222, 2/m2/m2/m, 2mm	$BaSO_4$, $PbCO_3$, K_2SO_4, Mg_2SiO_4
Monoclinic	1 twofold axis or 1 plane	(3), m, 2/m, 2	$CaSO_4 \cdot 2H_2O$, β-S
Triclinic	No axes or planes	(2), 1, $\bar{1}$	$CuSO_4 \cdot 5H_2O$, $K_2Cr_2O_7$
Hexagonal	1 sixfold axis	(7), $\bar{6}$, 6/m, $\bar{6}$2m, 6, 6mm, 622, 6/m2/m2/m	HgS, graphite, Mg, Zn
Rhombohedral	1 threefold axis	(5), 3, $\bar{3}$, 3m, $\bar{3}$2/m, 32	$NaNO_3$, As, Sb, Bi, α-quartz

[a] Number of groups noted in parentheses.

TABLE 2.3. Piezoelectric Point Groups

Crystal System	Full Hermann–Maugin Notation
Cubic	23
	$\bar{4}3m$
Tetragonal	4mm
	$\bar{4}2m$
	422
	4
	$\bar{4}$
Orthorhombic	222
	2mm
Monoclinic	m
	2
Triclinic	1
Hexagonal	$\bar{6}$
	$\bar{6}2m$
	6
	6mm
	622
Rhombohedral	3
	3m
	32

is called the *full Hermann–Mauguin* system of notation. As we have noted previously, for piezoelectricity to occur, the crystal must lack a center of symmetry. Accordingly, we can determine which point groups lead to materials exhibiting the piezoelectric effect (Table 2.3). Note that there are 20 piezoelectric point groups with at least one occurring in each crystal system. Note also for the cubic point group (432) that the threefold axis apparently reduces the piezoelectric effect to zero (6).

Piezoelectricity is also exhibited by materials which do not fall neatly into the "conventional" crystal systems we have described. For example, the effect is also seen in the synthetic fluoropolymer, poly(vinylidene fluoride), which belongs to the orthorombic crystal system, point group 2mm. This material has been used in a small number of acoustic wave sensor experiments. Layers of crystalline or polycrystalline materials can be deposited onto a non-piezoelectric substrate by molecular beam epitaxy or sputtering. In the case of polycrystalline films, one crystallographic axis of each crystal must be aligned in the growth process. As we shall see, ZnO has been deposited onto silicon substrates by such a process for application in the fabrication of acoustic wave sensors.

Thus far we have not indicated, at least in empirical terms, how single crystals from the piezoelectric point groups are actually processed before they can be used to construct devices. This leads us to a description of the plates cut from single crystals for this purpose. This is done to maximize the purity of a vibrational mode together with electromechanical and temperature-coefficient considerations for particular applications. Although we have no desire to confuse the reader, it is necessary to point out that two forms of notation (particularly for quartz) are employed for describing the crystal orientation of a selected plate. For quartz, Figure 2.2 depicts some of the various slices that have been used in acoustic engineering together with the older system of nomenclature, commonly known as AT-cut, ST-cut, and so on. The IEEE (7), however, has produced a standard notation which is based on a more rigorous description of the actual orientation of the plate with respect to the usual three orthogonal crystal axes, *X*, *Y*, and *Z*. For rectangular

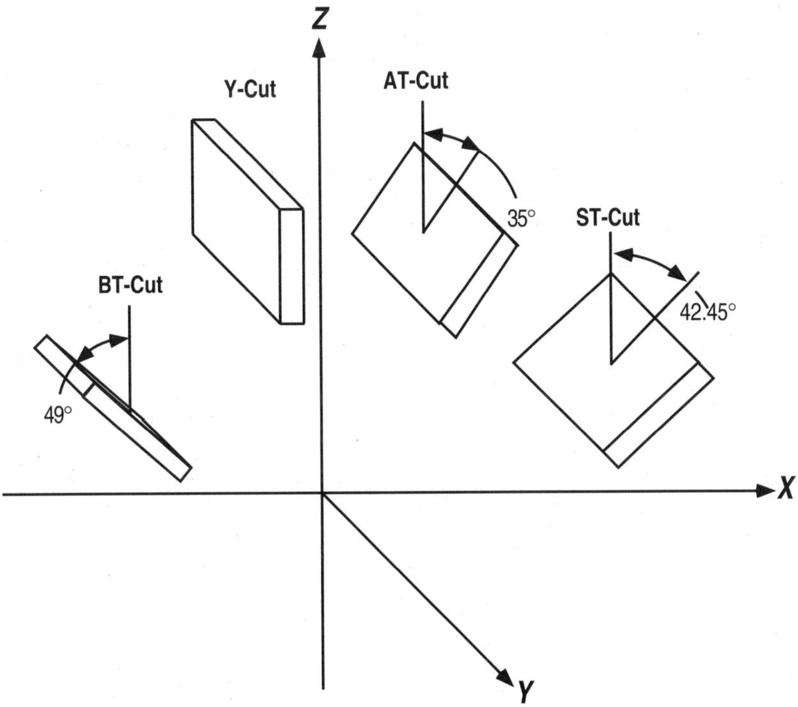

Figure 2.2. Typical cuts of quartz employed for piezoelectric applications shown relative to the crystal axes *X*, *Y*, and *Z*.

TABLE 2.4. Examples of Notations for Different Cuts of Quartz

Common Designation	Number of Rotations	IEEE Symbol
Nonrotated cuts	0	*XY, YZ*, and so on
AT	1	(*YXl*) −35°
BT	1	(*YXl*) 49°
GT	2	(*YXlt*) −51°/−45°
SC	2	(*YZwl*) 22.4°/−33.88°

plates the symbols *l*, *w*, and *t* represent the length, width, and thickness of the plate, respectively. These are used to denote the orthogonal axes fixed in a hypothetical reference plate, of which there are three. This reference plate is then subject to rotations about successive axes to achieve the final orientation.

A particular notation is defined by a symbol in which the first uppercase letter (*X*, *Y*, or *Z*) denotes the initial principle direction of the thickness, while the second denotes the initial principle length. The remaining letters (*l*, *w*, or *t*) describe about which of the three axes successive rotations were performed. Any symbol will have from two to five letters, depending on the number of required rotations. This is followed by a list of the successive rotation angles (Φ, Θ, Ψ) in strict order, separated by a solidus. By convention, angles are positive if the rotation is counterclockwise when viewed from the positive end of the axis toward the origin. Table 2.4 depicts the older nomenclature and IEEE standards for various cuts taken from a single crystal of quartz.

The mention of orthogonal crystal axes leads to an issue regarding the use of coordinate systems for the calculation of piezoelectric (see Section 2.2.2) and elastic constants. The material scientist or engineer contemplating incorporation of these constants in a theory of wave propagation must ensure that compability exists between a chosen coordinate system and the rectangular system outlined above. In addition, the relationship between the natural crystal axes, which form the basis of the *Miller* and *Miller–Bravais indices* systems, and the orthogonal crystal axes must be specified carefully. In other words, a practitioner possibly has to keep straight three different coordinate systems.

2.2.2. Coupling of Electromagnetism to Solid Elasticity

Our next task is to couple Maxwell's unifying laws of electromagnetism with the fundamental and material equations for elasticity introduced earlier. The end purpose is to obtain the particle motion with respect to time, that is, the wave equations for a piezoelectric material. Before embarking on the problem, there are three points to consider:

1. In general terms, full rigor with respect to the theory of electromagnetism is not required since the phase velocities of acoustic waves are in the region of four to five orders of magnitude less than the velocities of electromagnetic waves. This allows the ignorance of magnetic fields in a quasi-electrostatic approximation.
2. The external force $F_{m(j)}$ (Eq. 2.2) is usually set to zero.
3. The equations of the electric field are generally considered to be static, but in the piezoelectric model they are time dependent because of dynamical coupling to the mechanical equations.

Combination of the material equations with respect to stress and electrical field leads to the classical piezoelectric constitutive relations:

$$T_{ij} = c^{E}_{ijkl} S_{kl} + \eta^{E}_{ijkl} \dot{S}_{kl} - e_{ijk} E_{k} \tag{2.8}$$

$$D_{i} = \varepsilon^{s}_{ik} E_{k} + e_{ikl} S_{kl} \tag{2.9}$$

where E_k is the electric field, e_{ijk} is the third-rank piezoelectric stress coefficient tensor (18 possible values), ε^{s}_{ik} denotes the second-rank dielectric strain coefficient tensor (constant strain, six values), D_i is the electric displacement and the superscript E recognizes that the parameters c and η are in a constant electric field. The piezoelectric stress coefficient e_{ikl} is symmetric with respect to the indices l and k, while ε^{s}_{ik} is symmetric with respect to the indices i and k. The parameter D_i, which is a vector like E_i, perhaps needs some explanation for the chemist. It is, of course, well known that the application of an electric field can result in the formation of induced dipoles. This effect, which is called polarization, leads to the definition of the term $\boldsymbol{P}$, (for a dielectric medium) as the electric dipole measured per unit volume. It is proportional to the field $\boldsymbol{E}$ as follows:

$$\boldsymbol{P} = \chi \varepsilon_0 \boldsymbol{E} \tag{2.10}$$

where χ is often called the electric susceptibility by physicists and the polarizability by chemists. For an isotropic dielectric, the induced dipole moment is parallel to the electric field, so that χ takes a constant value for the material itself. For anisotropic materials, however, χ—like many of the parameters in this subject area—takes on the mathematical form of a tensor. The electric displacement, D_i, for a dielectric is now defined by

$$D_{i} = \varepsilon_0 E_{i} + P_{i} \tag{2.11}$$

where ε_0 is the permittivity of free space ($= 8.854 \times 10^{-12}$ F m^{-1}). In a sense, it represents the new local electric field after taking into account polarization effects.

The reader will now begin to see the link between the inherent time-dependant vectors such as E_i and D_i with particle motion through Equations 2.2, 2.8, and 2.9 (e.g., via T_{ij}). But before getting into time dependency and the wave equations, the physical significance of the coefficients mentioned above and a couple of new definitions are introduced. As we have noted above, the polarizability, χ—which is at the heart of piezoelectricity—can possess tensor character and therefore links P_i to E_i as follows:

$$P_i = \varepsilon_0 \chi_{ij} E_j + d_{ijk} T_{jk} \tag{2.12}$$

where d_{ijk} is called the third-rank strain coefficient tensor. Here, the term $\varepsilon_0 \chi_{ij} E_j$ is the familiar polarization produced by an electric field, while the second term is that generated by an applied stress. Stress, strain, and electric field are also linked as follows:

$$S_{ij} = d_{ijk} E_k + s^E_{ijkl} T_{kl} \tag{2.13}$$

where the first term is that part of the strain produced by the electric field while the second is that part produced by the applied stress. The coefficient s^E_{ijkl} is the fourth-rank compliance coefficient tensor under the condition of a constant electric field. Finally, we can demonstrate a link between electric displacement, electric field, and stress:

$$D_i = \varepsilon^T_{ij} E_j + d_{ijk} T_{jk} \tag{2.14}$$

where ε^T_{ij} represents the aforementioned second-rank dielectric strain coefficient tensor considered under constant stress. Being very careful with our notation, we can now draw the relationships between the various coefficients:

$$e_{ijk} = c^E_{ijkl} d_{klk} \tag{2.15}$$

$$\varepsilon^S_{ik} = \varepsilon^T_{ik} - d_{iij} c^E_{ijkl} d_{klk} \tag{2.16}$$

$$e_{ikl} = d_{iij} c^E_{ijkl} \tag{2.17}$$

and, by definition,

$$c^E_{ijkl} = (s^E_{ijkl})^{-1} \tag{2.18}$$

Equations 2.13 and 2.14 are rather important in characterizing the point groups with respect to crystals that we alluded to earlier in terms of the coefficients we have defined. In this regard, the equations have been expressed in a complete 9 × 9 *elastopiezodielectric* matrix:

$$\begin{bmatrix} S_1 \\ \cdot \\ \cdot \\ \cdot \\ S_6 \\ \\ D_x \\ D_y \\ D_z \end{bmatrix} = \left[\begin{array}{ccc|ccc} s^E_{11} & \cdots & s^E_{16} & d_{x1} & \cdot & d_{z1} \\ \cdot & & \cdot & \cdot & & \cdot \\ \cdot & & \cdot & \cdot & & \cdot \\ \cdot & & \cdot & \cdot & & \cdot \\ s^E_{61} & & s^E_{66} & d_{x6} & \cdot & d_{z6} \\ \hline d_{x1} & \cdots & d_{x6} & \varepsilon^T_{xx} & \cdot & \varepsilon^T_{xz} \\ \cdot & & \cdot & \cdot & & \cdot \\ d_{z1} & \cdots & d_{z6} & \varepsilon^T_{xz} & \cdot & \varepsilon^T_{zz} \end{array}\right] \begin{bmatrix} T_1 \\ \cdot \\ \cdot \\ \cdot \\ T_6 \\ \\ E_x \\ E_y \\ E_z \end{bmatrix} \tag{2.19}$$

This information can be summarized diagrammatically as shown in Figure 2.3, which takes as examples the monoclinic (m) and tetragonal (4) point groups. This exercise has, of course, been conducted for all crystal classes

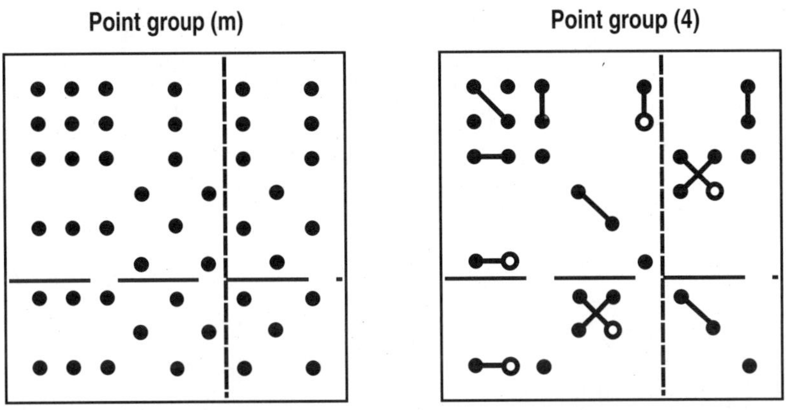

Point group (m)		Point group (4)	
Number of independent elastic constants	13	Number of independent elastic constants	7
Number of independent piezolectric constants	10	Number of independent piezolectric constants	4
Number of independent dielectric constants	4	Number of independent dielectric constants	2
	27		13

Figure 2.3. Examples of representation of elastopiezodielectric-matrices. A closed circle indicates a nonzero constant; an open circle indicates the negative of a closed circle ; a line between any two nonzero constants indicates numerical equality.

TABLE 2.5. Piezoelectric Stress Coefficients (e_{ijk}) and Dielectric Strain Coefficients (ε_{ik}) for Quartz

e_{ijk}	Value (Cm^{-2})	e_{ijk}	Value (Cm^{-2})	ε_{ik}	Value ($CV^{-1}m^{-1}$)
e_{111}	0.171	e_{222}	0.0	ε_{11}	39.21
e_{112}	0.0	e_{223}	0.0	ε_{12}	0.0
e_{113}	0.0	e_{233}	0.0	ε_{13}	0.0
e_{122}	−0.152	e_{311}	0.0	ε_{22}	39.82
e_{123}	0.067	e_{312}	0.067	ε_{23}	0.68
e_{133}	−0.0187	e_{313}	−0.0761	ε_{33}	40.42
e_{211}	0.0	e_{322}	0.0		
e_{212}	−0.095	e_{323}	0.0		
e_{213}	0.108	e_{333}	0.0		

and the resulting diagrams can be extremely useful in understanding the nature of piezoelectricity exhibited by a particular material. Finally, in this section we include some values for the piezoelectric stress and dielectric strain coefficients for quartz (Table 2.5).

2.2.3. Piezoelectric Wave Equations

The final job with respect to the theoretical aspects of piezoelectricity is to connect particle displacement, u_i, with the electric field, E_i, together with their time dependence. Our beginning point is Equation 2.2 from the discussion on elasticity, which is repeated here (minus the external force component):

$$T_{ij,i} = \rho_m \ddot{u}_j \tag{2.20}$$

We can substitute for the stress using Equation 2.8 and for the strain by Equation 2.3. Taking into account the symmetry arguments alluded to earlier, Equation 2.20 reduces to

$$-e_{ijk} E_{k,i} + c_{ijkl} u_{k,li} + \eta_{ijkl} \dot{u}_{k,li} = \rho_m \ddot{u}_j \tag{2.21}$$

From Maxwell's laws of electromagnetism and Equations 2.3 and 2.9, the following equation can be deduced:

$$E_{j,ij} - E_{i,jj} = -\mu_0 \left[\varepsilon_{ik} \ddot{E}_k + \frac{1}{2} e_{ikl} (\ddot{u}_{k,l} + \ddot{u}_{l,k}) \right] \tag{2.22}$$

where μ_0 ($= 4\pi \times 10^{-7}$ H m^{-1}) is the permeability of free space. The last two equations describe a system of coupled wave equations which associate u_i to E_i. Note that there are six scalar equations with six unknowns (u_1, u_2, u_3, E_1, E_2, E_3).

If we invoke the quasi-electrostatic approximation mentioned above, particle displacement can be related to electric potential. The electric field is the negative of potential gradient:

$$E_i = -\phi_{,i} \tag{2.23}$$

Where $\phi_{,i}$ is the gradient of the electric potential. Direct substitution for electric field in Equation 2.21 yields

$$e_{ijk}\phi_{,ki} + c_{ijkl}u_{k,li} + \eta_{ijkl}\dot{u}_{kli} = \rho_m\ddot{u}_j \tag{2.24}$$

Differentiating Equation 2.22 with respect to the coordinate x_i produces

$$\varepsilon_{ik}\ddot{E}_{k,i} + \frac{1}{2}e_{ikl}(\ddot{u}_{k,li} + \ddot{u}_{l,ki}) = 0 \tag{2.25}$$

The usual symmetry argument and substitution for the electric field yields

$$e_{ikl}\ddot{u}_{k,li} - \varepsilon_{ik}\ddot{\phi}_{,ki} = 0 \tag{2.26}$$

Integration of this equation gives

$$e_{ikl}u_{k,li} - \varepsilon_{ik}\phi_{,ki} = 0 \tag{2.27}$$

Thus, Equations 2.24 and 2.27 constitute a system of four scalar equations in four variables which relate u_i *to* ϕ.

We have now introduced, albeit briefly, the nature of possible waves propagating in linear piezoelectric materials. Clearly, this is a long way from the design and development of prospective devices, including those for use as chemical sensors. Accordingly, it is now important to optimize acoustic parameters such as the quality of electromechanical coupling and temperature stability through consideration of all the piezoelectric parameters discussed previously for a particular cut of a crystal. This exercise takes us into the realm of *transformation laws*, which deals with the conversion of one coordinate system to another (see Fig. 2.4). The reason for this procedure lies in the fact that the chosen coordinate system x, y, z employed above may not be the same as the rectangular X, Y, Z system used as described in determining

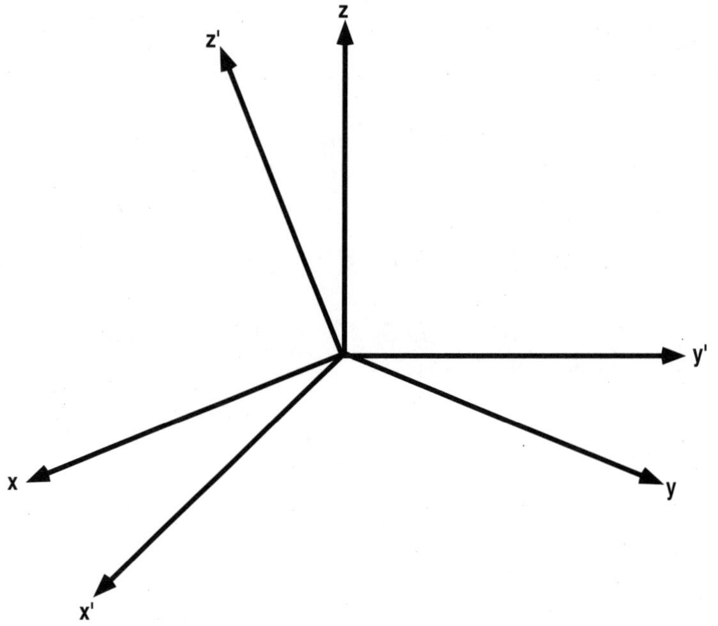

Figure 2.4. Example of old (x, y, z) and new (x', y', z') coordinate systems.

rotated crystal cuts. The necessary transformations are conducted by tensor mathematical procedures which are beyond the scope of this book, but are described in great detail elsewhere.

Finally, it is appropriate to comment in this section on the link between what we have considered so far and the structure of the various devices which will be introduced presently. With piezoelectric materials, we possess the ability to create acoustic waves by applying oscillating electrical fields and the ability to detect those waves by measuring electrical fields. This is generally achieved by depositing metal electrodes on the surface of the chosen slice of crystal in a very particular orientation. Application of the theories outlined above can be used to calculate the properties of the material when it is subjected to an electric field of a specific geometry. Such calculations are, however, extremely complex and, therefore, to satisfy the goals of this book we concentrate for the most part on an empirical description of induced wave motion. The direct employment of these devices as chemical sensors involves, of necessity, the additional introduction of boundary conditions to solve the nature of particle motion as it is affected by an added film.

2.2.4. Resonance

Readers who appreciate the music emanating from string instruments will be familiar with the phenomenon of resonance. It is a very important consideration in the design and fabrication of acoustic wave devices employed for a number of applications. In this section we give an empirical description of the phenomenon, which is expanded in subsequent chapters.

We start with the elementary physics of what happens when an acoustic wave, such as a uniform plane wave, propagates through a medium and is incident upon a boundary separating the matrix from a different medium. As is the case for light waves, reflected and transmitted (refracted) waves may be involved. Snell's Law and the Fresnel equations will be familiar to the scientist concerned with optical devices, and these ideas are just as applicable to acoustic wave physics. When acoustic waves are travelling in a solid phase–piezoelectric or nonpiezoelectric–the nature of the second medium is crucially important. Snell's Law describes the direction of scattered waves at a boundary, whereas the Fresnel equations take into account wave amplitudes. These equations can be solved for various pairs of media, although the process is quite complex and beyond the scope of this book, especially for piezoelectric materials.

The connection between boundary effects and resonance is that a structure that is finite in size and functioning in air will, to a first approximation, confine the vibrations–in piezoelectric or nonpiezoelectric solids–to the device. For a piezoelectric sensor, one can view the motion of the material in the form of standing waves which must satisfy both mechanical and electrical boundary conditions. In an oscillating electric field the crystal is vibrating which, in turn, sets up electric charges on the device surface that interact with this field. Accordingly, a resonance can be instigated which invokes the *natural frequency of oscillation.* In an empirical view one can think of this process as involving complete constructive interference of the standing acoustic waves mentioned above. A device operating under these conditions is termed a *piezoelectric acoustic resonator.*

2.3. PARTICLE DISPLACEMENT AND ACOUSTIC WAVES

We are now approaching a description of the basic structure, including the nature of wave propagation, in a selection of devices that have been employed for the purposes of chemical sensing. Before accomplishing this task it is necessary to introduce the basics of solid material deformations (or vibrations) that can vary with time. This involves expressing the most elementary

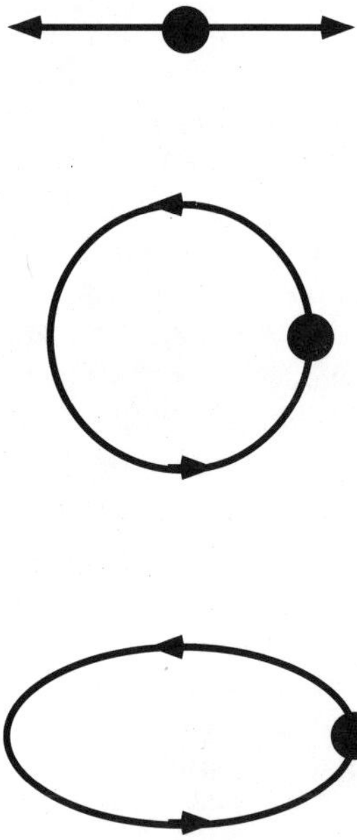

Figure 2.5. Schematic representation of particle (●) trajectories in linear, circular, and elliptical polarizations.

components of the general wave equations—Equation 2.7 for nonpiezoelectric materials and Equation 2.24 for piezoelectric solids—in a simple pictorial form.

Displacement of particles from their equilibrium positions by a stress together with the effect of elastic restoring forces leads to travelling or standing waves in the solid material of interest. For a sinusoidal vibration with time, there are three possible geometries of motion for a specific particle. The actual movement is called a *polarization.* The possibilities (Fig. 2.5) are linear, elliptical, and circular polarizations. The various vibrational modes that can be propagated in solids can be constructed from these different polarizations. There are, however, two basic patterns of deformation that can occur, namely, the *shear* and the *compressional* wave. To appreciate the nature of these patterns, it is quite instructive to visualize them through grid diagrams. In Figure 2.6 we depict an arbitrary three-dimensional grid which is at rest and not strained, together with a grid for particles which are subject to movement or polarization in one plane only, the z plane. Note that the wave motion is actually in the y direction, as shown in the diagram. This motion constitutes a shear wave in that the wave propagation is perpendicular to the particle displacement. It is often called a *transverse* wave. This is a definition we will return to presently. A full description of the vibration depicted in

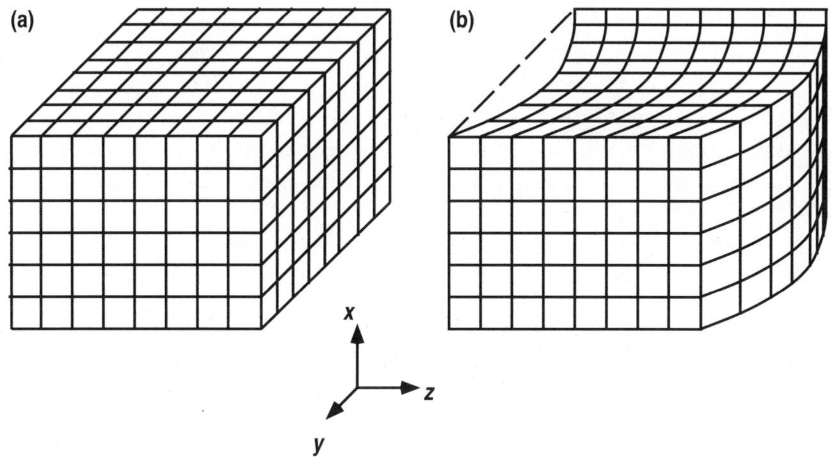

Figure 2.6. Grid diagram for (*a*) at-rest solid and (*b*) z-polarized, y-propagating shear wave.

Figure 2.6 would be a *z-polarized, y-propagating shear wave.* Finally, we should note that propagation of shear waves in a cubic crystal will clearly result in symmetric orthogonal vibrations which are then called *degenerate* shear waves.

The compressional wave can also be represented by a grid diagram, and this is given in Figure 2.7. In this case, the particle motion is in the same

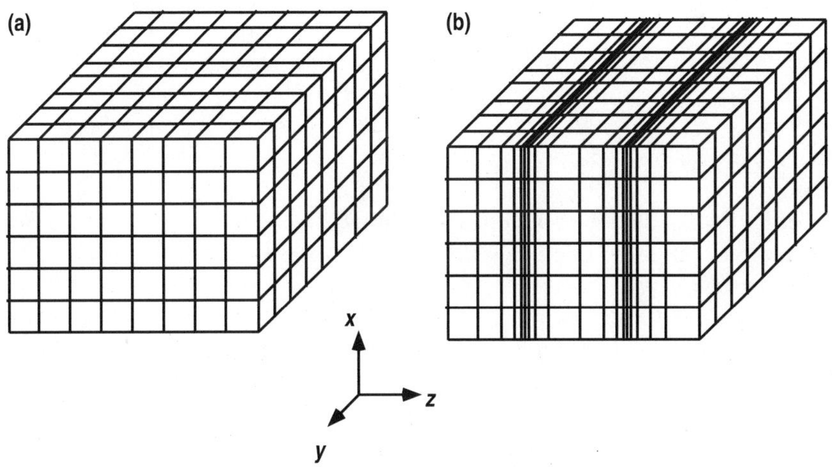

Figure 2.7. Grid diagram for (*a*) at-rest solid and (*b*) z-polarized compressional wave.

direction as the wave propagation and involves particles with decreasing (compression) and increasing (rarefaction) proximity to each other. The compressional motion is often called a *longitudinal* wave. The wave shown in Figure 2.7 is simply termed a *z-polarized compressional wave.* Finally, we should note that for a motion of particles in one plane only, shear or compressional, the convention has it that the vibration is called an *acoustic plane wave.*

2.4. ACOUSTIC-WAVE DEVICES EMPLOYED TO STUDY INTERFACIAL CHEMISTRY

As we have alluded to earlier, acoustic wave devices have found wide-ranging application in electrical and electronic engineering. In this section, we concentrate only on the structure of those devices that have been employed for sensing chemistry at the device surface and, in particular, those structures that involve the launching of surface waves. There is some minor confusion as to what constitutes a surface-wave device, so we shall attempt to delineate carefully the nature of the waves propagating in the sensor. This is obviously going to be crucial in terms of the interaction of the acoustic energy with the chemistry imposed at the device–gas or device–liquid interface.

The nature of the acoustic wave generated in piezoelectric materials is determined by the crystal cut (and therefore orientation) and thickness of the material used, as well as the geometry and configuration of the metal electrodes employed to instigate the electric field. For nonpiezoelectric substrates, both the method of exciting vibrational modes together with the efficiency of coupling of acoustic energy to the sensor structure *and* the geometry and acoustic impedance of the device will be important.

Finally, all of the devices for which descriptions follow have been used specifically as chemical or biological sensors. This implies that a chemically selective layer is conventionally added to the surface of a particular device in order to bind an analyte from the gas or liquid phase. Accordingly, an important consideration will be the analytical sensitivity compared to what might be termed the physical sensitivity of the device. The latter involves the actual perturbation of the device wave velocity, amplitude, and/or frequency by the presence of the analyte in the chemically selective layer, while the former is the first derivative of the measured response *versus* the *sample concentration.* Great care must be taken in evaluating these two forms of sensitivity when contemplating use of a device as a chemical sensor or instrument for the study of thin-film chemistry. Subsequent to introduction of the various devices, we

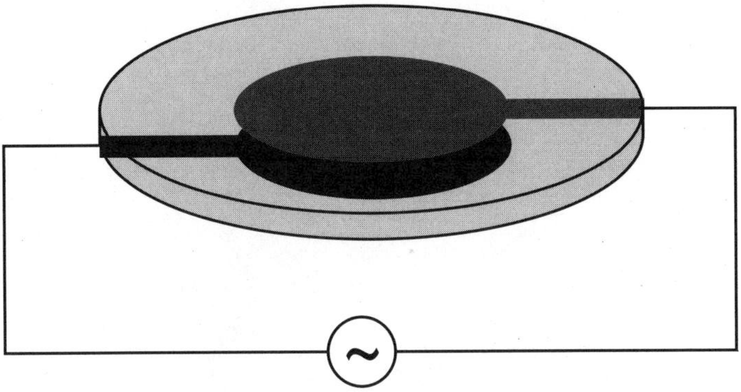

Figure 2.8. Typical TSM sensor with electrode connections.

shall return to a brief treatment of this issue. Furthermore, the question of analytical sensitivity is the subject of additional discussion later in the text.

2.4.1. The Thickness-Shear Mode Sensor

As discussed in Chapter 1, this device was the first to be introduced. The description of the device as a thickness-shear mode (TSM) sensor is applied in a generic fashion to emphasize the fact that bulk shear waves are travelling in a piezoelectric substrate. When the substrate is quartz, the sensor has often been referred to as a quartz crystal microbalance (QCM). This terminology alludes to the observation that the frequency response of the device can sometimes be correlated with surface mass changes. Given the modern trend of using substrates other than quartz, and of using acoustic devices to study interfacial physical properties, the term QCM is perhaps outmoded.

The construction of a typical TSM sensor is shown in Figure 2.8. The piezoelectric substrate is a thin disc of AT-cut quartz (see Section 2.2.1) and the oscillating electrical field is applied employing two metal transducers (electrodes) on opposite sides of the wafer. The electrodes are typically sputtered onto the quartz and can be composed of gold, silver, aluminum, palladium, or other metals. Electrically speaking, the structure constitutes a *one-port* (two-terminal) device.

The acoustic waves propagate in a direction perpendicular to the particle displacement, which is parallel to the disc surface in Figure 2.9. Because there is no particle movement in the y and z directions, the wave motion is

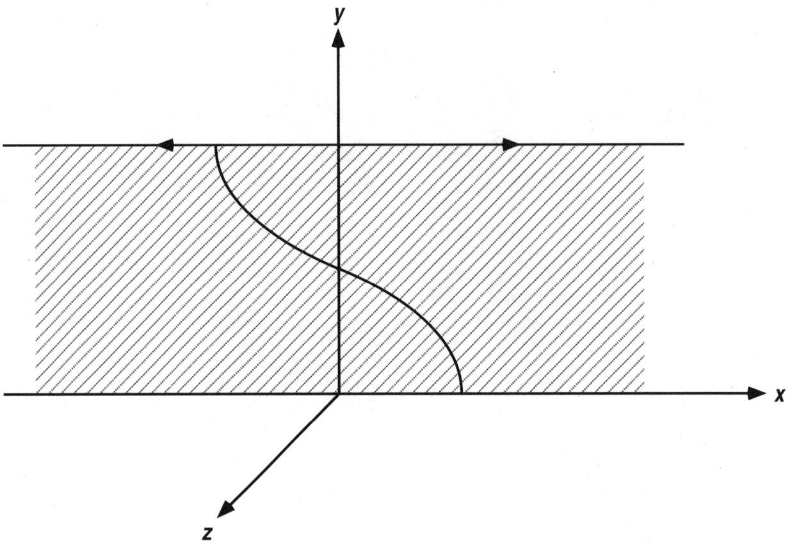

Figure 2.9. Schematic of thickness dimension of a TSM wafer depicting particle movement to generate a y-propagating shear wave.

called shear or transverse and we can easily relate the motion to the wave equations (Eqs. 2.24 and 2.26) developed previously.

$$c_{1212}\frac{\partial^2 u_x}{\partial y^2} + \eta_{1212}\frac{\partial^3 u_x}{\partial y^2 \partial t} + e_{212}\frac{\partial^2 \phi}{\partial y^2} = \rho_m \frac{\partial^2 u_x}{\partial t^2} \tag{2.28}$$

$$e_{212}\frac{\partial^2 u_x}{\partial y^2} - \varepsilon_{22}\frac{\partial^2 \phi}{\partial y^2} = 0 \tag{2.29}$$

Note that we have related the single direction of particle displacement, u_x, to the electric potential, ϕ, and that the various coefficients apply to quartz as does the density term, ρ_m. We must be careful to recognize that these differential wave equations relate to the coordinate system applied to the actual disc of quartz. To place these equations in the context of the crystallographic X, Y, and Z axes referred to in Section 2.2.1 requires the use of the transformation laws mentioned previously.

The device is almost always operated in the resonance condition (see Section 2.2.4). This means that constructive interference of standing waves takes place when taking into account the fact that the waves are reflected from the two surfaces of the disc. Accordingly, the plate thickness, t, in the y

direction yields the wavelength, λ, of the fundamental ($n = 1$) and harmonic ($n = 3, 5, 7 \ldots$) resonances through the simple expression

$$\lambda = \frac{2t}{n} \tag{2.30}$$

The practical limitation of working with quartz plates of convenient thickness (of the order of 10^{-4} m) means that the resonant frequency for $n = 1$ is in the region of 5–15 MHz (the frequency increases as the thickness decreases).

Finally, it is instructive to note that particle motion at the electrode surface is parallel to the plane of the added chemically selective layer.* This feature presents particularly interesting possibilities for the study of interfacial molecular species at the sensor–liquid junction. The motion of the device generates a damped wave as it propagates into this medium. Accordingly, chemistry conducted at the interface can perturb the transmission of acoustic energy which is being transferred to the liquid and, in turn, lost due to viscous forces. This results in the possibility of examining surface roughness, free energy, and viscosity.

2.4.2. Surface-Launched Acoustic Wave Devices

Before embarking on a consideration of the various structures, we must reiterate the definition and concept that lies behind the term *surface launched.* The definition here is that the wave motion is excited by transducers imposed at the surface of the substrate, and that the movement is detected by the same or an additional transducer *on the same surface.* As we shall see, this can mean that a device can fall into the "surface-launched" category even though bulk waves are propagating in the substrate (because the thickness of a particular device is of the same order as the acoustic wavelength). One additional point to note is that the use of the word "surface" here is not the one familiar to the physical chemist who works with the likes of scanning-tunnelling microscopy or X-ray photoelectron spectroscopy. Surface wave to the engineer translates to confined motions found in a zone "near" the surface of a solid. Accordingly, the terminology means quite different things to the two communities.

*Although the electrode is not piezoelectric, it is forced into motion by the instigated oscillating movement of the quartz plate. Therefore, in the simple configuration described here, the necessary chemistry must be conducted on the electrode surface.

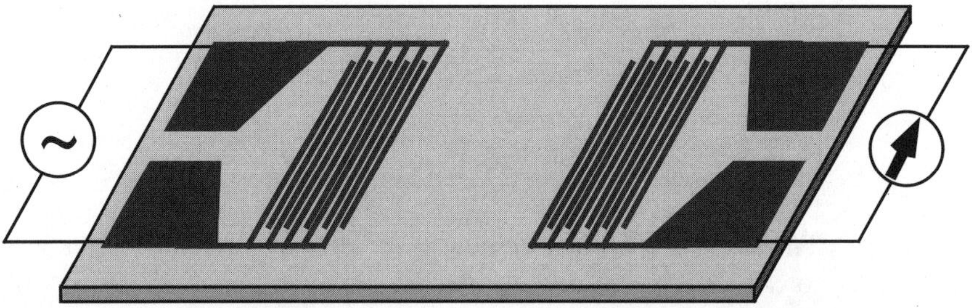

Figure 2.10. Typical Rayleigh-wave acoustic device with interdigital "finger" electrodes to generate and receive acoustic waves.

We now turn to the first device to be considered in this category–the *surface acoustic wave* (SAW) or Rayleigh wave sensor. In this structure, an interdigital transducer (IDT) fabricated from a metal is deposited on a plate of ST-cut quartz (for example) which is of a thickness much greater than the acoustic wavelength (λ in this case is of the order of tens of micrometers). Figure 2.10 depicts a schematic of the "fingers" used for generating Rayleigh waves in the SAW structure. The waves propagate in both directions from the transducer and are confined to approximately one acoustic wavelength from the piezoelectric substrate surface. This confinement is a result of the plate thickness and wavelength being such that both the electric field and corresponding particle displacements decay to zero within the top layers of the plate.

Before introducing the nature of the Rayleigh wave, we need to describe briefly the existence of two types of acoustic shear vibrations, namely, the shear vertical (SV) and shear horizontal (SH) waves. In this case, the terms vertical and horizontal are specified with reference to a particular boundary plane. Figure 2.11 depicts two hypothetical single waves propagating in an arbitrary slice of a solid slab. As labelled, the vertical axis (y) refers to the thickness of the slab and the horizontal axis (x) to the length dimension. The SV wave involves particle motion in the y direction, whereas the SH wave is comprised of particle motion in the z direction. It is important to note that in both cases the wave is propagating in the x direction, hence they are referred to as shear or transverse waves. The reference plane mentioned above is the top view of the slab depicted in Figure 2.11.

We now turn to the nature of the Rayleigh wave instigated by the IDT. For this type of wave propagation, particles in the surface region of a piezoelectric material possess elliptical motion. The wave consists of both compressional and shear components, which are coupled by the boundary surface, as de-

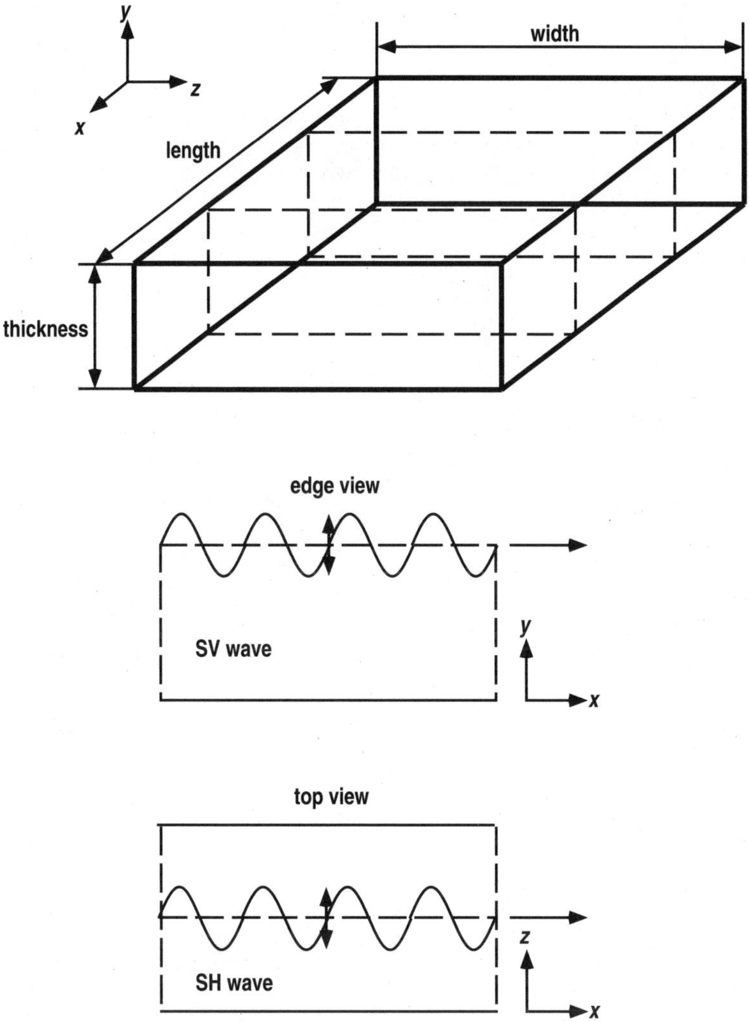

Figure 2.11. Distinction between shear vertical (SV) and shear horizontal (SH) waves.

picted in Figure 2.12 (to simplify the representation only one plane is shown). Note that the shear wave is of the y-polarized, z-propagating type and can be categorized as an SV wave based on the definition specified above. The compressional wave is clearly a z-polarized, z-propagating wave. With respect to the coordinate axes given in Figure 2.12, the plane containing the

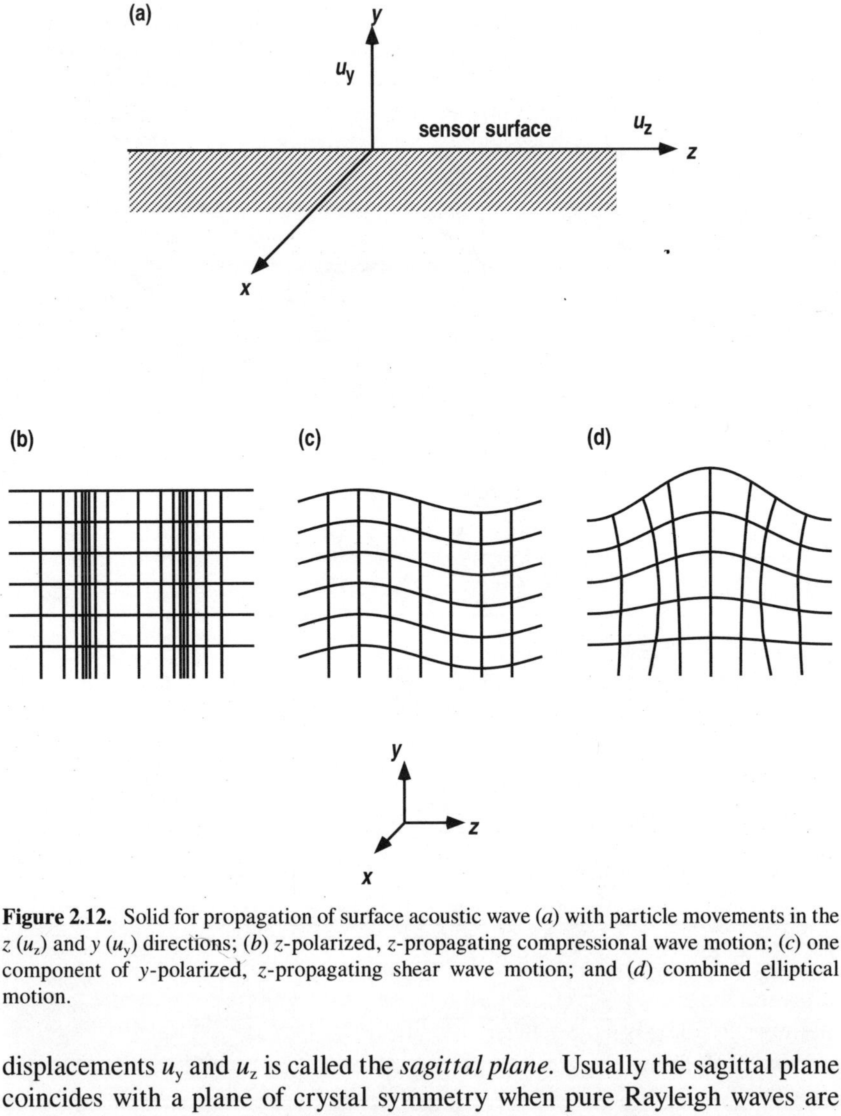

Figure 2.12. Solid for propagation of surface acoustic wave (*a*) with particle movements in the z (u_z) and y (u_y) directions; (*b*) z-polarized, z-propagating compressional wave motion; (*c*) one component of y-polarized, z-propagating shear wave motion; and (*d*) combined elliptical motion.

displacements u_y and u_z is called the *sagittal plane.* Usually the sagittal plane coincides with a plane of crystal symmetry when pure Rayleigh waves are propagating in a material. Finally, we note that solution of the wave equations (Eqs. 2.24 and 2.26) for Rayleigh wave propagation in a particular class of piezoelectric material is a very difficult task. Indeed a number of approximations and numerical methods must employed.

The SAW device is generally operated in two different ways. In the first structure, IDTs are placed at opposite ends of a slice of piezoelectric material.

One IDT generates a surface acoustic wave which propagates to the ouput transducer, where it is detected. This is called a *delay-line* configuration because of the transit time associated with the much lower velocity of the acoustic wave compared to the much faster propagation of electromagnetic energy. The delay-line structure is called a *two-port* device. In the second configuration, either one or two IDTs are arranged such that surface waves propagate in one direction and are then reflected back to create a resonant cavity. Typical frequencies of operation for these devices lie in the range 20–300 MHz. Finally, with respect to these devices, we emphasize that the rather general acronym SAW is almost always used when referring to the pure Rayleigh wave sensor. Other devices, as we shall now see, also involve surface waves, but the various structures are given different names and, therefore, acronyms.

We now turn to two devices that have their genesis in a reduction of the thickness of the Rayleigh wave device to a dimension of the order of the acoustic wavelength. The structures to be outlined here are called *plate wave* devices. More specifically they are referred to as *flexural plate wave* (FPW) and *shear horizontal-acoustic plate mode* (SH-APM) sensors. They are considered to be surface-wave devices because the acoustic energy is instigated in a fashion somewhat analogous to a Rayleigh wave. It should be emphasized, however, that the waves actually propagate through the bulk of the sensor.

The FPW device was introduced by White and co-workers in the 1980s (9). A typical structure (Fig. 2.13) is constructed following procedures characteristic of those used for fabricating microelectronic devices. An orientation-dependant etchant such as ethylene diamine/pyrocatechol is used to remove Si from the reverse-side of a silicon wafer (the substrate). This process can be controlled to produce a thin membrane which is of the order of micrometers in thickness. The generation of the silicon membrane may be followed by depostion of thin metal films such as chromium or gold (for adhesion purposes), followed by RF planar magnetron sputtering to yield a layer of piezoelectric ZnO. The IDTs can then be formed on the ZnO layer by conventional photolithography. It is important to realize that the plate or membrane thickness (about 3 μm) used in this type of device is less than the acoustic wavelength, which leads us to the topic of the generation of *Lamb waves*.

A brief discussion of Lamb waves begins with the concept of waveguiding. Waveguides—which will be very familiar to the electromagnetic radiation specialist—are also very important in the propagation of acoustic energy, particularly with respect to the fabrication of delay-line devices. Acoustic energy is propagated in the direction of the guide as is the case with optical waveguides. It should be noted that wave equations are generally derived for

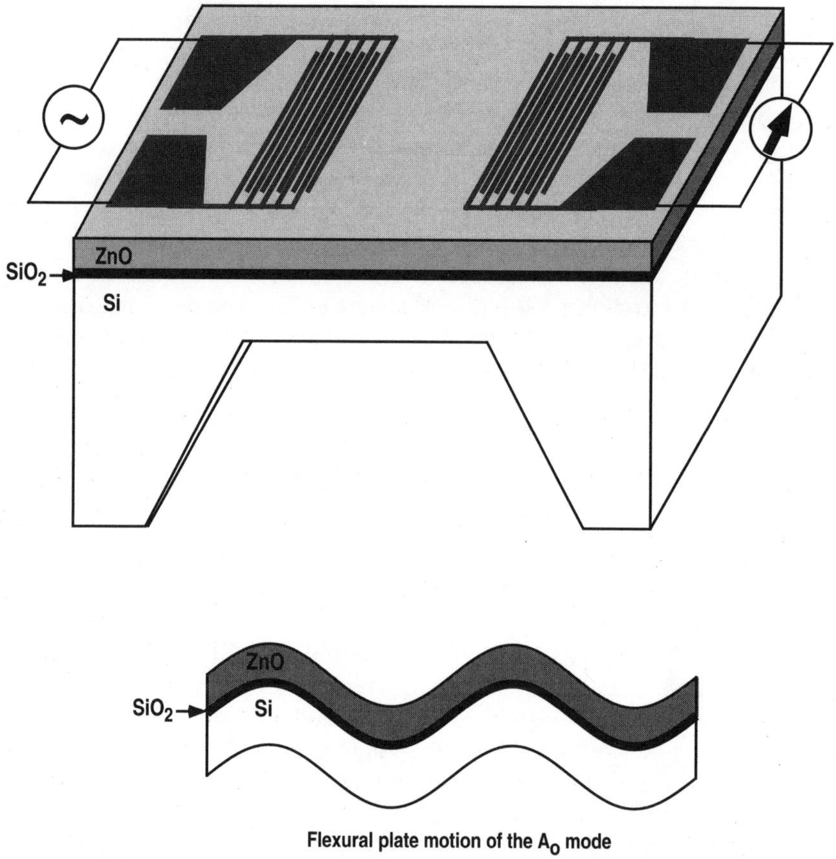

Figure 2.13. Flexural plate-wave sensor and cross section showing plate motion. Note depostion of ZnO on a thin silicon membrane.

propagation through infinite or semiinfinite substrates. Waveguides are layers or structures having dimensions of the order of the acoustic wavelength. Therefore, solution of the wave equations (Eqs. 2.24 and 2.26), taking into account electrical and mechanical boundary conditions for a piezoelectric thin plate or film, leads to the *waveguide modes.* Although not used as a waveguide, the FPW device shown above displays analogous physics where a modal analysis can be performed.

In a plate of the thickness mentioned previously, acoustic vibrations can be instigated by the presence of an electric field associated with an IDT. These waves are called Lamb waves. The particle displacements for these waves, as with the Rayleigh wave, are elliptical. Also, the motion has components which are longitudinal and SV in form, with the latter being transverse to the direction of wave propagation. In addition, the waves can be divided into two categories, that is, the symmetric (S) and antisymmetric (A) modes, where the term symmetry is defined by the particle displacement about the median plane of the plate (Fig. 2.14). As would be expected, the two modes also possess several orders of wave motion, with the lowest order vibrations often being termed S_0 and A_0. In the symmetric modes, the boundaries of the plate dilate and contract periodically, whereas in the antisymmetric modes there is a periodic flexing motion; accordingly, these two modes are called *dilatational* and *flexural,* respectively (Fig. 2.14). Finally, the FPW device is arranged with respect to electrical IDT excitation such that only flexural waves are propagated in the device. For an interesting theoretical treatment involving solution of wave equations 2.24 and 2.26 together with models of

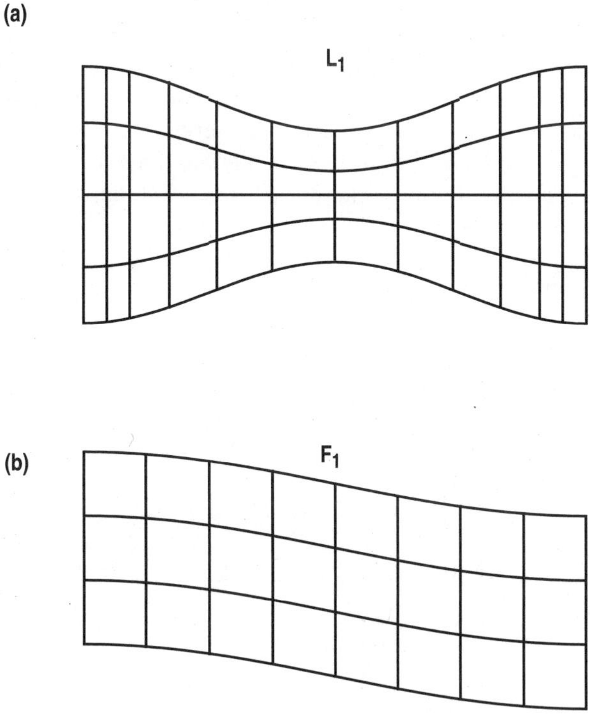

Figure 2.14. Grid representation of particle displacements for (*a*) symmetric (dilatational) and (*b*) antisymmetric (flexural) Lamb waves for the $n = 1$ modes.

electrode-bounded and nonbounded plates, the reader might consult the excellent paper by Joshi and Jin (10).

The FPW device is claimed to posses several advantages over other sensors, particularly its Rayleigh wave sister. It can be operated at much lower and convenient frequencies ($\approx$5 MHz), the structure lends itself to microfabrication methods (like the SAW), and the device can be used for applications in both gas and liquid phases. The latter is associated with the fact that the velocity of the flexural wave is significantly lower than the compressional velocity of sound in liquids. Accordingly, the device does not generate compressional wave energy in liquids and, therefore, is not severely attenuated. We shall revisit application of the FPW as a liquid phase sensor in later chapters.

We now turn to the second plate-mode device—the SH-APM structure. (The reader must be careful with nomenclature here, because the device described immediately above could quite easily be designated as an "APM" device). This sensor was introduced by Martin and co-workers (11) during the same time period as the aforementioned FPW device. The structure of a typical SH-APM device is shown in Figure 2.15. As for the SAW device, the SH-APM sensor begins with a plate of ST-cut quartz upon which are deposited input and output IDTs (a two-port device format). The plate thickness is of the order of 200 μm, and the IDT fingers are designed such that the sensor operates at frequencies in the approximate range 100–150 MHz. Note from Figure 2.15 that the particle motion here is composed entirely of dis-

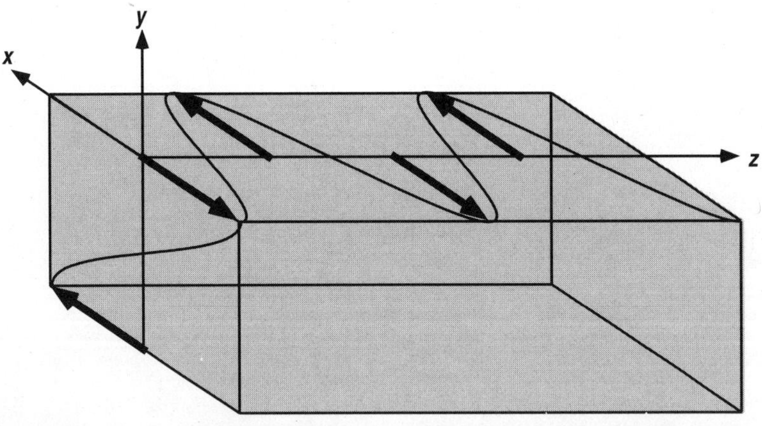

Figure 2.15. Shear horizontal acoustic plate mode (SH-APM) device depicting the particle displacement for the $n = 1$ mode.

placements within the xz-plane (see earlier discussion of SH waves). Additionally, as for the FPW sensor, there are several modes with different orders, that is, $n = 0, 1, 2, 3$, and so on. The sensor has been thought of as an acoustic waveguide, with SH plane waves being subject to superposition by multiple reflection between the upper and lower faces of the quartz plate. A particular mode of interest can be resolved by choice of plate thickness in comparison with transducer periodicity. The sensor can be used in both gas and liquid phases, the latter being associated with the fact that motion of surface quartz particles is parallel to the sensing surface. Accordingly, like the TSM device described earlier, coupling to the liquid is essentially associated with viscous forces in the liquid. The SH-APM offers the additional advantages (over the TSM) of higher mass sensitivity and ability to perform chemistry (and place the sample) on the reverse side of the device to that where the transducers are formed. Again, applications of the SH-APM will be discussed later.

The final structure we will introduce in the surface-launched category of acoustic wave sensors, the surface transverse wave (STW) device, is a relative newcomer to the field (12). In a sense, it constitutes an attempt to maximize some of the advantages of the various devices we have described thus far. Before describing the structure of the sensor, it is useful to point out that structures in which thin films are used to guide (see above) acoustic waves are often called Love wave devices. In the STW, acoustic waves are "trapped" in the surface zone of the device by a metal grating which is formed between the transducers. The device, shown in Figure 2.16, is fabricated on an underlying substrate of 0.5 mm thick ST-cut quartz. The IDTs and grating fingers are 5-μm wide aluminum, with a period of 20 μm and a thickness of 200 nm. The waveguide is 10 mm long and the entire surface is coated with 50 nm of SiO_2 to protect the electrodes from corrosion. In an innovative step, the SiO_2 also provides an ideal surface for attaching a chemically selective layer. The frequency of operation of the device is in the region of 250–500 MHz.

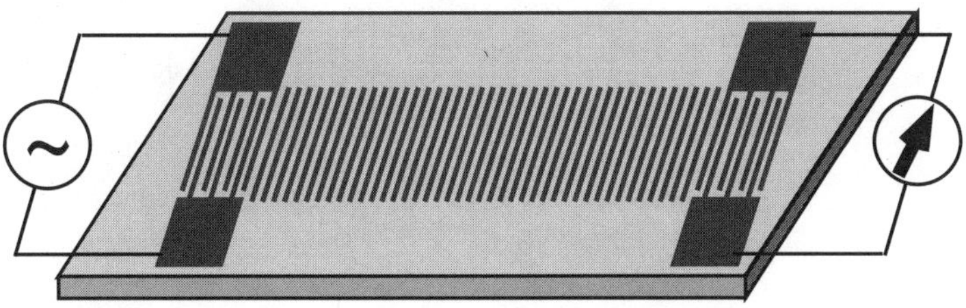

Figure 2.16. Surface transverse wave device. Note that the substrate ends are cut at an angle to the long axis of the IDTs.

Finally, as shown in Figure 2.16, the particle motion is in a similar direction to that of the SH-APM device, with the acoustic wave being transverse to this axis. The difference between the STW and SH-APM is that the acoustic energy is said to be genuinely confined to the surface in a "pure" shear horizontal mode. This feature apparently leads to greater analytical sensitivity.

2.4.3. Nonpiezoelectric Acoustic Wave Devices

As we indicated in our discussion of solid elasticity, it is possible to propagate acoustic waves in nonpiezoelectric materials, although this approach has not figured prominently in chemical sensor development. In particular, we will describe here the relatively new thin-rod acoustic wave (TRAW) sensor that was introduced by Jen and co-workers (13) . This device has been exploited as a chemical sensor by Thompson and Li (14). The experimental arrange-

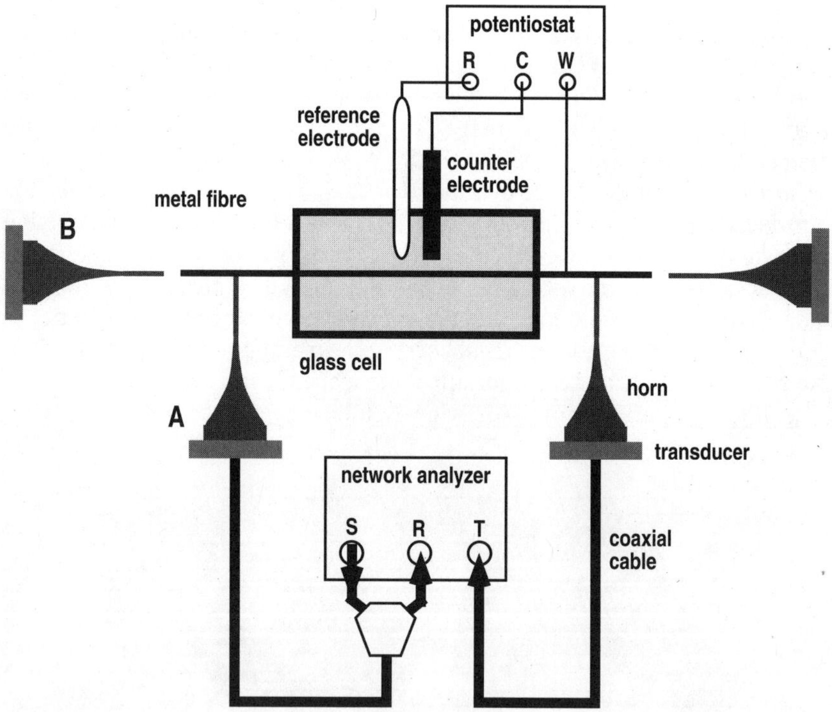

Figure 2.17. Thin-rod acoustic wave sensor configuration with arrangements for inducing flexural (A) and extensional (B) modes of vibration.

ment in its most basic form is depicted in Figure 2.17. An alternating current of radio frequency is fed into a piezoelectric input transducer which generates an ultrasonic longitudinal wave. The latter is associated with particle movement perpendicular to the transducer surface. This longitudinal vibration then induces a glass horn, which is acoustically coupled to the transducer, to also vibrate. The glass horn, in turn, is connected to a thin rod or fiber composed of a nonpiezoelectric material in a 90° or 0° arrangement. The thin rod must propagate an acoustic wave with wavelength much greater than the radius of the device. As indicated already, however, the fiber can be fabricated from any material of choice because propagation of the acoustic wave is based entirely on mechanical processes. The whole structure can be generated in the familiar delay-line configuration which has been described previously.

Depending on the nature of excitation of vibrational energy in the fiber, three types of oscillation can be induced—the *torsional, extensional* and *flexural modes:* as examples, we consider briefly the latter two modes, which have actually been investigated from an analytical chemical point of view. The extensional mode involves no circumferential movement and, like the SAW device described earlier, is composed of two coupled modes, the axial, L_{oi} and radial, R_{oi}, motions. An attempt to depict the particle displacements is given in Figure 2.18. Both axial and cross-sectional representations are shown. The flexural mode is similarly displayed in the same figure. In the latter case excitation is such that the transducer is connected at 90° to the fiber.

The structure, which in its present form could hardly be described as a device, possesses several advantages over some of the sensors described above. The thin rod can be constructed from a pure metal which allows its use as an electrode in electrochemical research. In addition, the structure can be operated at low frequencies ($\approx$200 kHz) with the same analytical sensitivity as a TSM, and in liquids in the flexural mode.

Finally, the TRAW sensor concept has been extended to a tube geometry (15). In this case, the wall thickness of the tube is much smaller than the acoustic wavelength. So far, only the extensional wave has been propagated in a tube structure. This device allows the rather exciting possibility of performing chemical or biosensor measurements directly on flowing liquids (through the tube).

2.4.4. Brief Comparison of Devices

Since a number of the devices described above will only be revisited from an analytical chemical point of view in a limited fashion, a short comparison of their properties follows. A comparison of some of the features of the acoustic

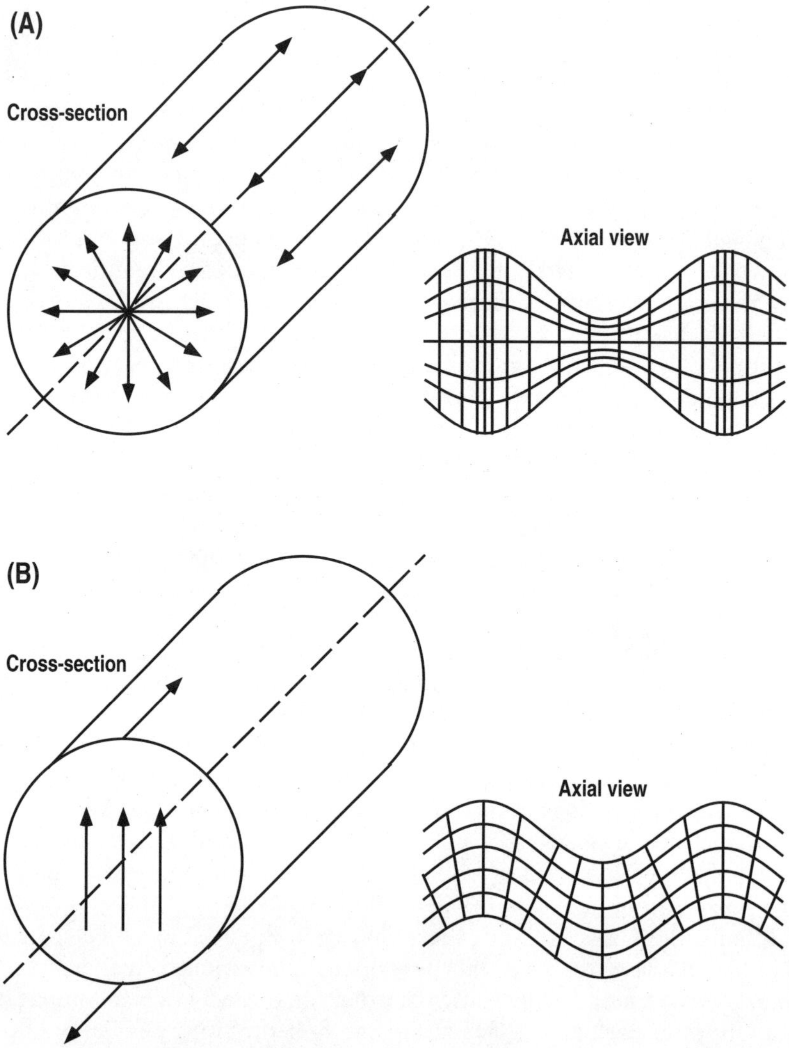

Figure 2.18. Particle displacements for the lowest-order (A) flexural and (B) extensional modes of vibration in TRAW sensors.

wave devices discussed are given in Table 2.6. Note that the only device which is inoperable in the liquid phase is the SAW sensor, although there was an earlier indication that this might be possible.

A particularly important parameter for evaluating the various structures is the sensitivity, in classical analytical chemical terms, of the sensor to species

TABLE 2.6. Characteristics of Acoustic Wave Devices

Device	Wave Launched	Particle Displacement Components[a]	Transducers	Operating Frequency (MHz)	Analytical Phase	Mass Sensitivity[b]
TSM	Bulk	Transverse	Planar Electrode	5–15	Gas, liquid	$-\frac{1}{\rho_m d}$
SAW	Surface	Transverse Parallel	IDT	30–500	Gas	$-f_0\frac{K_1}{\rho_m V}$
FPW	Surface	Transverse Parallel	IDT	2–7	Gas, liquid	$-\frac{1}{2\rho_m d}$
SH-APM	Surface	Transverse	IDT	25–200	Gas, liquid	$-\frac{J}{\rho_m d}$
STW	Surface	Transverse	IDT	200–500	Gas, liquid	$-f_0\frac{K_2}{\rho_m V}$
TRAW	Bulk		Piezoelectric Horn	0.2–0.5	Gas, liquid	$-\frac{L}{\rho_m a}$

[a] Relative to direction of wave propagation.

[b] Key: ρ_m-density of sensor material; d, plate thickness; V, acoustic wave velocity; f_0, unperturbed device frequency; K_1, constant depending on plate properties; K_2, constant depending on plate properties and grating; a, radius of cross section of thin rod; $J = 0.5$ for $n = 0$ and $J = 1$ for $n > 0$; $L = 1$ for first extensional mode and $L = 1/2$ for first flexural mode. Both FPW and SH-APM expressions include the free isotropic plate assumption.

present in the analytical matrix under study. Generally, in analytical chemistry, instrumental sensitivity is defined as

$$R = f([\mathrm{A}]) \tag{2.31}$$

where R is some form of response and [A] is the concentration of analyte species A. Differention of this equation will yield the sensitivity which may be, for example, a linear, log, or power function of concentration. In reality, it is quite difficult to express such a relationship for the devices considered so far, since they all function on the principle of analyte being bound at the device surface. Important factors that are expected to contribute to the overall response of the sensor will be the phase that the analyte is distributed in, transport-to-sensor capture dynamics, and the selectivity of the surface "trapping" process for the analyte. Perhaps a more direct strategy for comparing the sensitivity of the various devices is the response of the device once the analyte is *bound* on the sensor surface (a rather "unreal" analytical chemical situation!). This is by no means a straightforward issue either, since the sensor might respond to changes in surface film chemistry associated with viscoelasticity and modulus. We return to all of these considerations in more quantitative terms in later chapters.

Despite the problems presented by the factors mentioned above, we can perform a comparison of the devices based on their sensitivity to bound mass, although such a concept might be considered of limited usefulness. In an analogous fashion to Equation 2.31, the mass sensitivity of the devices, $S_{\mathrm{m}}^{\mathrm{V}}$ (based upon phase velocity measurement at constant frequency), is defined by

$$S_{\mathrm{m}}^{\mathrm{V}} = \lim_{\Delta M \to 0} \frac{1}{V} \frac{\Delta V}{\Delta M} \tag{2.32}$$

Note that this expression contains the incremental change in phase velocity, ΔV, relative to the actual velocity, V, for an incremental change in the added mass per unit area. If V_0 and V are the phase velocities before and after mass deposition, respectively, and $V \approx V_0$, the following approximation can be used:

$$S_{\mathrm{m}}^{\mathrm{V}} = \lim_{\Delta M \to 0} \frac{1}{V_0} \frac{\Delta V}{\Delta M} \tag{2.33}$$

The direct relationship between velocity and frequency can be used to generate the following approximation (although it must be pointed out that Equation 2.33 is given for constant frequency):

$$S_m = \lim_{\Delta M \to 0} \frac{1}{f_0} \frac{\Delta f}{\Delta M} \tag{2.34}$$

Note that this quasi-absolute expression for mass sensitivity has the units of $cm^2\ ng^{-1}$ if the added mass is given in nanograms and the unit area in square centimeters. However, to yield an idea of the change in frequency caused by added material, the units are often given as, for example, (Hz/MHz)/(ng/cm^2).

In very general terms, it would be expected that the expressions for S_m in Table 2.6 for the various devices can be subject to comparison for measurements conducted in the gas phase, where most of the acoustic energy is confined to the device and any added mass. The liquid phase, however, is a different matter altogether. In this case, the expressions themselves possess little validity, so comparisons become meaningless. For the gas phase, we can see from Table 2.6 that the mass sensitivity of the four devices in which the whole structure is subject to vibration (TSM, SH-APM, FPW, and TRAW) is increased as the thickness of the device decreases. As we described above, this is in a sense optimized with respect to the FPW device, which is based on a very thin plate or membrane. One might consider that reduction of the sensor thickness of the TSM and SH-APM structures and radius of cross section of the TRAW device could result in increased sensitivity, but the practical limitation becomes the fragility of the plate or fiber. Finally, with respect to these four devices one might believe that there is a general relationship between S_m and frequency, because of the connection between device dimension and wavelength. This is not the case, however, for the FPW and TRAW (flexural mode) sensors where reducing the aforementioned dimension results in lower frequency.

The mass sensitivity of the two remaining devices, the SAW and STW, is determined by the extent to which the Rayleigh and transverse waves penetrate into the piezoelectric substrate from the real surface. This depth is decreased for reduction of wavelength and for increased grating thickness in the SAW and STW devices, respectively.

In the following chapters, we concentrate entirely on the surface-launched devices, with heavy emphasis on the SAW because of its prominence in the field of analytical chemistry.

REFERENCES

1. W. F. Cady, *Piezoelectricity,* Volumes 1 and 2, Dover, New York, 1964.
2. V. E. Bottom, *Introduction to Quartz Crystal Unit Design*, Van Nostrand-Reinhold, New York, 1982.

3. H. F. Tiersten, *Linear Piezoelectric Plate Vibrations (Elements of Linear Theory of Piezoelectricity and the Vibrations of Piezoelectric Plates),* Plenum Press, New York, 1989.
4. B. A. Auld, *Acoustic Fields and Waves in Solids,* Volumes 1 and 2, Wiley, New York, 1973.
5. V. M. Ristic, *Principles of Acoustic Devices,* Wiley, New York, 1983.
6. Reference 5, p. 182.
7. *IEEE Standard on Piezoelectricity,* Institute of Electrical and Electronic Engineers, New York, 1988, *ANSI/IEEE Std. 176-1987.*
8. J. W. Grate, S. J. Martin, and R. M. White, *Anal. Chem.,* **1993,** *65,* 940A.
9. C. T. Chuang, R. M. White, and J. J. Bernstein, *IEEE Electron Dev. Lett.,* **1982,** *EDL-3,* 145.
10. S. G. Joshi and Y. Jin, *J. Appl. Phys.,* **1991,** *70,* 4113.
11. S. J. Martin, A. J. Ricco, T. M. Niemczyk, and G. C. Frye, *Sens. Actuators,* **1989,** *20,* 253.
13. C. K. Jen, J. E. B. Oliveira, J. C. H. Yu, J. D. Dai, and J. E. Bussiére, *Appl. Phys. Lett.,* **1990,** *56,* 2183.
14. P. C. H. Li and M. Thompson, *Anal. Chem.,* **1996,** *68,* 2590.
15. P. C. H. Li and M. Thompson, *Anal. Chim. Acta,* **1996,** in press.

CHAPTER

3

THEORY OF ACOUSTIC WAVE SENSOR RESPONSE

In Chapter 2 we introduced the various device formats currently employed in piezoelectric chemical sensor technology. In this chapter we discuss the theory of SAW chemical sensors in terms of various parameters affecting sensor response. As already mentioned, the device most commonly refered to as a SAW sensor is more accurately referred to as a Rayleigh wave device. Since the majority of the published theoretical studies concern this particular device, however, we adopt here the literature convention that the generic term "SAW" implies a Rayleigh wave device *unless* specified otherwise. We also take the somewhat unusual step of describing the response of surface-launched acoustic wave devices to chemical and/or physical perturbation of the device–gas or –liquid interface *before* providing a detailed description of interdigital transducer function. The reason for this is that the IDT itself constitutes a "coating" on the piezoelectric substrate, and certain of the principles outlined in this chapter are key to the explanation of IDT function. Thus, while we present a descriptive treatment here, a more rigorous treatment of the latter follows in Chapter 4 in the context of SAW sensor-system design and operation.

3.1. BASIC THEORY OF OPERATION

The successful development of the interdigital transducer (IDT) by White and Voltmer (1) allowed direct piezoelectric coupling to surface acoustic waves. Prior to this, generation and detection of such waves could only be achieved through mechanical coupling. An IDT consists of a pair of overlapping, interspersed "finger" arrays (Fig. 3.1*a*) deposited onto a suitable substrate. Normally there are an equal number of fingers on each half of the IDT, with the finger width equal to the finger spacing (a 1 : 1 metalization ratio). The cut of the substrate and orientation of the IDT with respect to the crystallographic axes are chosen such that when a potential is applied across the IDT, the substrate undergoes a periodic deformation (Fig. 3.1*b*). When an alternating potential is applied, an acoustic wave is launched which propagates away

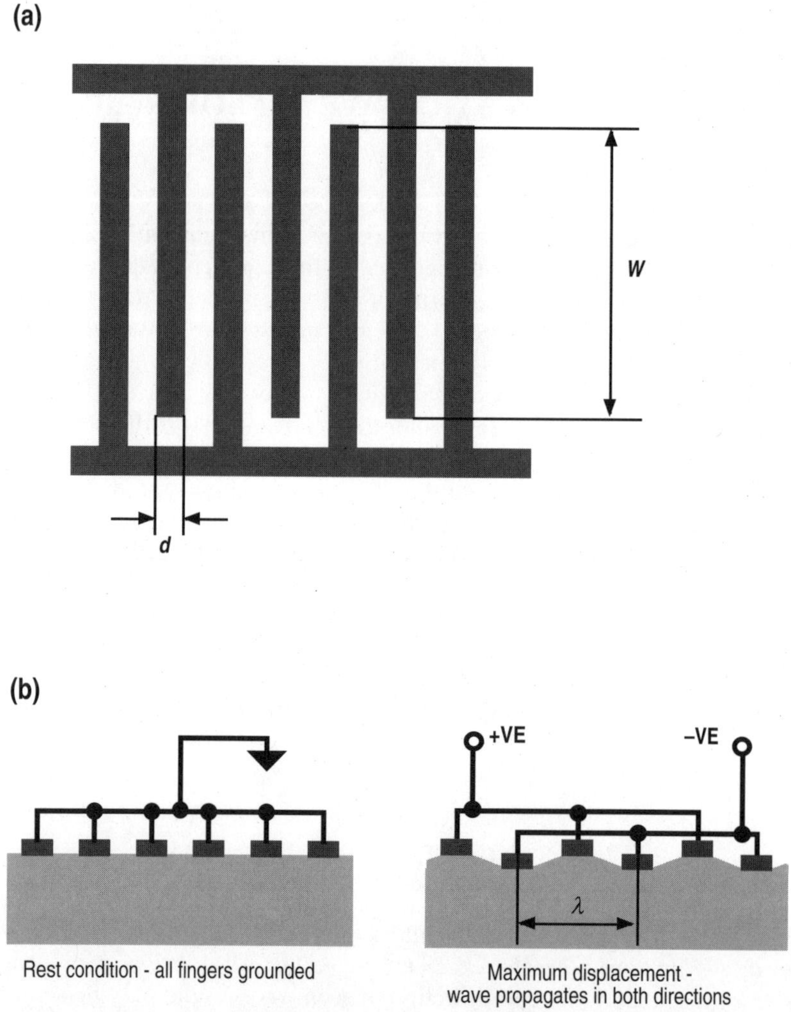

Figure 3.1. (*a*) Schematic of a typical IDT, showing the finger width (d) and bandwidth (W). (*b*) Launching of a SAW by applying a potential across alternating fingers of an IDT.

from the IDT in both directions and perpendicular to the long axis of the IDT fingers. The acoustic wavelength will then be determined by the IDT geometry and spacing d as defined in Fig. 3.1*a*:

$$\lambda = 4d \tag{3.1}$$

For a typical substrate such as ST-quartz, with acoustic propagation along the *X*-crystallographic axis, it can be shown that the resulting acoustic wave is a Rayleigh wave whose amplitude decays exponentially into the bulk of the substrate (2). This type of wave has both shear horizontal and vertical components, so that the motion of any particular point in the path of the wave describes an elliptical orbit about its rest position (Section 2.4.2). Since the acoustic wave-propagation velocity will be determined by the acoustic mode, orientation, and substrate, the fundamental resonance frequency of the device is simply

$$f_0 = v_0/\lambda \tag{3.2}$$

where f, v, and λ are the wave frequency, acoustic propagation velocity, and wavelength, respectively, and the subscript 0 indicates unperturbed conditions. For a given acoustic mode, any perturbation of the propagation velocity can be measured as a change in the resonance frequency, because the wavelength is fixed by the geometry of the IDT. Furthermore, changes in v will result in a changed phase difference between the applied electric potential and that measured at a fixed point along the propagation path from the launching IDT. It is this combination of resonant frequency and phase shift that has made SAW devices of considerable importance as electronic components in high-frequency circuits (3–5).

The key to SAW chemical sensors, then, is understanding which factors can affect the acoustic propagation velocity and, therefore, result in changes of phase and resonance frequency. Following the notation of Ricco et al. (6), these can be expressed as

$$\frac{\Delta v}{v_0} = \frac{1}{v_0}\left[\frac{\partial v}{\partial m}\Delta m + \frac{\partial v}{\partial c}\Delta c + \frac{\partial v}{\partial \varepsilon}\Delta \varepsilon + \frac{\partial v}{\partial \sigma}\Delta \sigma + \frac{\partial v}{\partial T}\Delta T + \frac{\partial v}{\partial p}\Delta p + \cdots\right] \tag{3.3}$$

where m is mass, c is stiffness, ε is dielectric constant, σ is surface conductivity, T is temperature, and P is pressure. It is possible to obtain expressions for $\Delta v/v_0$ for a variety of different cases using a perturbational approach and the appropriate boundary conditions (7). In terms of the chemical sensor literature, attention was initially focused on the effect of added surface mass since the first SAW devices all employed a thin polymeric, metallic, or metal oxide layer deposited on the propagation path. The influence of the sheet conductivity of these films was also recognized at an early stage of development, while physical parameters such as temperature and pressure were considered largely from the perspective of eliminating unwanted system noise. Continuing work has examined in greater detail the many factors which can make a substantial contribution to observed sensor response. In the following sections, we examine each of these in more detail.

3.2. PHYSICAL PARAMETERS AFFECTING SAW SENSOR RESPONSE

3.2.1. Effect of Operating Temperature

It is well known that temperature can have a considerable affect on propagation velocity for acoustic wave devices, since many of the material constants involved are themselves temperature dependent. The temperature dependence of surface wave attenuation for quartz, in particular, has been studied over various temperature and frequency ranges (8–10). Different piezoelectric materials, propagation directions, and cuts all show different temperature dependencies over different ranges. It is thus possible to engineer a SAW device to have specific temperature characteristics, which has led to the development of several SAW-based temperature sensors (11–13).

To illustrate, Hauden et al. (10) have expressed the effect of temperature in terms of the corresponding frequency shift relative to f_0 at 20°C, and compared experimental measurements with calculations based on the known temperature coefficients of the elasticity constants for both ST- and Y-cut quartz. For ST-X quartz, f_0 shows a square dependence on the temperature centered at 20°C. In fact, the ST cut was developed because it showed little or no temperature dependence at room temperature. The first- and second-order temperature coefficients derived by Hauden et al. are $-0.24 \times 10^{-6}\,°C^{-1}$ and $-40 \times 10^{-9}\,°C^{-2}$, respectively. Table 3.1 gives calculated Δf

TABLE 3.1. Calculated Frequency Shifts due to Temperature for a 52-MHz ST-X Quartz SAW Delay Line

T (°C)	$-\Delta f$ (kHz)
20	0.00
30	0.33
50	2.25
100	14.31
150	36.77
200	69.64

values for a 52-MHz ST-X quartz SAW device. Temperature is of particular relevance to chemical sensors because the kinetics of sensor response are themselves temperature dependent.

3.2.2. Effect of Applied Pressure

Pressure (gas loading) also contributes to acoustic wave attenuation and velocity change, and has been studied extensively by Slobodnik (14). In this case, the attenuation is due to the generation of compressional waves in the gas in contact with the SAW device (Fig. 3.2). In other words, the shear vertical component of the wave causes periodic compression and rarefaction of the gas, resulting in a coupling of acoustic energy from the device into the gas. The condition for this to occur is given by Slobodnik as $\cos\zeta = \lambda/\lambda_f$ (see Fig. 3.2). Clearly, this process will be more effective the greater the density and pressure of the gas, and so the greater will be the attenuation. In fact, Slobodnik's results show that the attenuation is reasonably approximated by the ratio $(\rho_f v_f)/(\rho_s v_s \lambda)$ where subscript "f" indicates the fluid (gas) loading the substrate (subscript "s"). For example, it was found that attenuation increased with atomic mass for the noble gases, levelling off at Kr. In addition, attenuation was found to vary linearly with pressure. The SAW devices can, therefore, also be used as pressure sensors, as demonstrated by Risch (15).

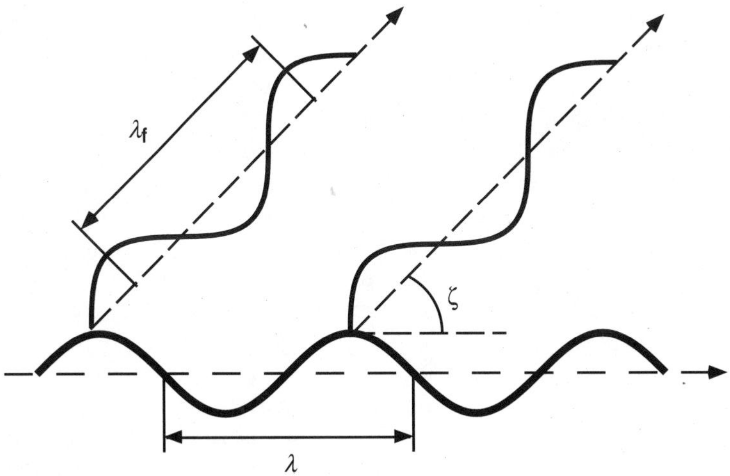

Figure 3.2. Generation of a compressional wave into a fluid (λ_f) by a SAW (λ) at an angle ζ to the device surface.

Other acoustic propagation modes also show pressure dependence, as demonstrated by Wenzel and White (16) for a Lamb wave sensor exploiting the lowest-order antisymmetric flexural plate mode. Such devices show a strong velocity perturbation on fluid loading, which may also be accompanied by amplitude attenuation if the acoustic velocity in the fluid is less than that for the substrate. For the lowest antisymmetric mode, Wenzel and White derived the relationship:

$$v = k\sqrt{\frac{B}{M + (\rho_f K/k)}} \tag{3.4}$$

where k is the wave number ($= 2\pi/\lambda$), B is a stiffness term, M is the mass per unit area of the membrane, ρ_f is the fluid density, $K = [1 - (v/v_f)^2]^{1/2}$, v_f is the fluid acoustic velocity and the term (K/k) is the decay length of the evanescent wave propagating into the contacting fluid. Thus, the sensor response varies with the fluid density (and hence pressure). Results for a 4.74-MHz delay-line oscillator show that the variation of frequency shift with absolute pressure (applied to both sides of the membrane) can be approximated by a linear function over limited ranges, while the response to differential pressure (pressure drop across the membrane) is linear as determined by liquid loading experiments. Other applications of the FPW device exploiting equation 3.4 are discussed in Chapter 6.

3.2.3. Effect of Applied Electric Potential

In Chapter 2 we described how an electric potential can cause a mechanical deformation through the piezoelectric effect. One implication of this is that any acoustic wave propagating through a piezoelectric substrate will have an accompanying electric potential wave arising from the alternating mechanical separation of charges. In the presence of a conducting or semiconducting film, this gives rise to various effects, as described below. It also means that any charge applied, or in proximity, to the surface of a SAW device can give rise to a change in the acoustic propagation velocity. This latter property has been exploited by various workers to produce voltage sensors (17–19).

As with other effects, the substrate material, cut, and propagation direction are all important in determining the linearity and sensitivity of the frequency response to applied voltages. Inaba et al. (17) used a piezoelectric ceramic as the substrate. This gave a nonlinear frequency response for electrostatic potentials applied via a coupling capacitor. Both Joshi (18) and Gatti et al. (19) report good linearity for Y-X $LiNbO_3$ with high applied d.c. voltages. It should be noted that in all three examples the device requires two electrodes (signal and ground) applied to opposite faces of the SAW delay line in the propagation path. Thus the conductivity and thickness of the electrode on the propagating surface also play a key role in determining the sensitivity and zero-voltage frequency (Section 3.5).

3.3. INFLUENCE OF SURFACE MORPHOLOGY

Surface morphology can have a profound affect on acoustic wave propagation and, subsequently, on device properties. Isolated protuberances or surface defects, and periodically or randomly rough surfaces, can all contribute to a loss of surface wave energy and changes in impedance characteristics. Note that this roughness may be due to the physical nature of a real surface (device surfaces are typically optically flat with a root-mean-square deviation of the order of 10 nm or less) or from structures deposited on the surface such as IDTs, periodic gratings, and isolated strips. The latter cases are of particular importance in the design of SAW filters (3–5,7).

In essence these effects are due to scattering of the Rayleigh waves, which may also result in conversion, for example, to Lamb waves (7). Early efforts to theoretically quantitate these effects focused on perturbational analysis of sinusoidally and sawtooth-roughened surfaces (20). For sinusoidal roughening of the surface, Rayleigh wave attenuation is a function only of the Poisson ratio of the substrate, and shows two maxima when plotted as a function of the period of the surface roughness. Attenuation is increased for decreasing values of the Poisson ratio.

A theoretical study of the acoustic properties of more realistic surfaces is clearly a complex task; however, it is one that is attracting increasing attention (21). For example, Maradudin (22) has considered the properties of an isolated protuberance, a periodic grating, and a randomly rough surface. Calculations showed that in the first case, acoustic resonances occur at discrete frequencies localized in the vicinity of the protuberance. For typical acoustic propagation velocities and protuberance widths of 10 μm, these resonances are of the order of 100 MHz. Maradudin further showed that surface waves propagating across periodically corrugated surfaces can be converted into bulk waves. This leads to increased energy loss and a decrease in the propagation velocity relative to a perfectly planar surface, effectively reducing device-resonant frequency while certain frequencies will not be propagated at all (see also Section 3.7.3).

Similar properties occur for a randomly rough surface, which can be thought of as a series of superposed gratings having different orientations with respect to the direction of propagation in the surface plane. This will cause Rayleigh wave attenuation through scattering and conversion to BAWs, and is believed to be the most important extrinsic mechanism for attenuation.

As is the case for a periodic grating, wave slowing will occur, while some acoustic modes can be supported that would not otherwise exist. Maradudin's calculations show that increasing surface roughness will reduce f_0, except at some critical value where there is zero shift in frequency.

Thompson et al. (23) have reported studies in which the propagating surface of SAW delay lines was etched with ammonium bifluoride in an attempt to increase the active surface area for vapor adsorption. The devices used in this study were 52-MHz ST-X quartz delay lines ($\lambda = 60\ \mu$m). For short etch times producing features of $\approx 1\ \mu m^2$ area, there does appear to be the predicted decrease in f_0. For longer etch times giving etch features of $\approx 100\ \mu m^2$, however, f_0 appears to increase. Quite large variations in f_0 were observed simply for remounting the SAW device in the test apparatus, however, so it is not clear whether this latter result is a real effect due to increased propagation by other acoustic modes or simply an experimental artefact.

3.4. EFFECT OF AN ELECTRICALLY INSULATING SURFACE FILM

So far we have only considered the effect of external physical forces. In all practical chemical sensors, however, some surface coating is necessary to impart the required degree of chemical sensitivity and selectivity. This then adds to the number of factors that can perturb v_0. Much effort has been applied to identifying these factors and putting their effects onto a sound theoretical basis. Different types of surface layer give rise to differing degrees of complexity in the derived expressions for the resulting frequency shifts. Following the perturbational analysis previously mentioned (7), Wohltjen (24) showed that the frequency shift obtained for a thin, nonpiezoelectric, lossless, insulating isotropic film using the Tiersten boundary conditions (25) is

$$\Delta f = (k_1 + k_2) f_0^2 h \rho_f - k_2 f_0^2 h \left\{ \frac{4\mu'(\lambda' + \mu')}{v_0^2(\lambda' + 2\mu')} \right\} \tag{3.5}$$

where Δf is the frequency shift, k_1 and k_2 are substrate material constants, h is film thickness, ρ_f is film density, and λ', μ' are the film first and second Lamé constants (c_{12} and c_{44} from the stiffness matrix), respectively. Both λ' and μ' are also identified as the bulk and film shear elastic modulii, respectively. Note that since k_1 and k_2 are both negative in sign, Equation 3.5 implies increasing mass loading (first term) will cause a frequency *decrease,* while increasing shear modulus (second term) will cause a frequency *increase.* Values of k_1 and k_2 for various piezoelectric materials are given in Table 3.2. An exact solution, while more complex, is possible (25); the values

TABLE 3.2. Material Constants for Different Piezoelectric Substrates Used in Equation 3.5[a]

Substrate	v_0 (m/s)	k_1 ($\times 10^{-8}$ m^2s/kg)	k_2 ($\times 10^{-8}$ m^2s/kg)
Y-X quartz	3159.3	−9.33	−4.16
ST-X quartz	3158	−8.7	−3.9
Y-Z $LiNbO_3$	3487.7	−3.775	−1.73
Z-X or Y ZnO	2639.4	−5.47	−2.06

[a] Values taken from references 24 and 27.

predicted by the perturbation method deviate increasingly from the exact solution with increasing film thickness. Values of h for which Equation 3.5 are valid are such that $h \ll \lambda$. A solution may also be derived for anisotropic surface films (7).

Wohltjen has pointed out that the term $(\lambda' + \mu')/(\lambda' + 2\mu')$ is algebraically constrained to lie between 0.5 and 1.0. Further, under certain conditions, the right-hand term of Equation 3.5 becomes negligible, reducing to

$$\Delta f = (k_1 + k_2) f_0^2 h \rho_f \tag{3.6}$$

These conditions are met when μ' is low ($\approx 10^6$ Pa), which corresponds to values typical of soft, rubbery polymers such as poly(isoprene). For rigid, glassy materials ($\mu' \approx 10^9$ Pa), however, the second term becomes significant and can actually give rise to a frequency increase for film deposition (Table 3.3).

TABLE 3.3. Theoretical Effect of Mass Loading and Shear Modulus on Δf for Different Film Coatings on a 52-MHz ST-X Quartz SAW Device

μ' (Pa)	$\dfrac{\lambda' + \mu'}{\lambda' + 2\mu'}$	Δf (kHz)
1.00×10^9	1.00	+2.52
	0.50	+1.25
1.00×10^8	1.00	+0.233
	0.50	+0.106
1.00×10^7	1.00	+0.0049
	0.50	−0.0078
1.00×10^6	1.00	−0.018
	0.50	−0.019

Attempts to verify Equations 3.5 and 3.6 have primarily involved Langmuir–Blodgett deposition of various large organic species or coating with different polymeric materials (Table 3.4) (24,26–31). (Note at this point that we are concerned only with the effect of the surface coating, not with vapor sorption into that coating—this will be dealt with separately in Section 3.7.) In all cases listed in Table 3.4, a frequency *decrease* was observed with increasing film thickness. In some cases, however, this was nonlinear, while the slope was systematically lower than predicted by theory.

TABLE 3.4. Experimental Verification of Equation 3.5

f_0 (MHz)	λ (μm)	Coating	h (μm)	Layers	Linear	Ref.
31	102	Poly(methyl methacrylate)	2.1	—	√	24
		Poly(isoprene)[a]	0.75	—	×	24
		Poly(ethylene maleate)	2.0	—	×	24
		Unspecified LB film[b]		45	√	27
		Stearyl alcohol/CuPC		45	√	28
52	61	Tetrakis(cumylphenoxy)phthalocyanine[c]	0.004	65	√	26
		Unspecified LB film[b]		45	√	27
		Stearyl alcohol/CuPC		45	√	28
		Polycarbonate	0.5	—	×	65
98.6	32	Tetracyanoquinodimethane[d]		20	×	30
		Butyl silicon phthalocyanine dichloride[e]		11	√	31
		ω-Tricosenoic acid[e]		15	√	31
112	28	Unspecified LB film[b]		45	×	27
158	20	Unspecified LB film[b]		45	√	27
		Poly(vinyltetradecanal)[e]		49	√	29
200	15	Poly(vinyltetradecanal)[e]		69	√	29
300	10	Poly(vinyltetradecanal)[e]		31	√	29
400	7.9	Poly(vinyltetradecanal)[e]		19	√	29
434		Poly(vinyl chloride)	?	—	√	80
		Bisphenol-A/carbenium ion	?	—	√[f]	80

[a] Response lower than predicted with increasing h.

[b] Slope lower than predicted.

[c] Weakly conducting ($\approx 10^{-10}$ S cm^{-1}).

[d] Conducting film with linear mass component.

[e] Insulating film.

[f] Slope only 25–50% of that predicted by mass effects.

This latter point can be explained in part by the lack of shear modulus data for the films examined. This is especially important for SAW experiments, since measurements of shear modulii are frequency dependent. In one case, the deviation was due to the contribution of film conductivity (30). To the best of our knowledge, no one has ever reported an increase in f_0 for film deposition alone, although this is theoretically possible for a highly rigid, glassy material (Table 3.3).

3.5. INFLUENCE OF A CONDUCTING SURFACE FILM

An acoustic wave propagating through a piezoelectric medium is accompanied by an electric potential wave (ϕ). This wave can be affected by an applied electric field or the properties of a conducting or semiconducting surface layer. One case of particular relevance to SAW devices is the short-circuit condition obtained by applying a thin conducting layer to the surface, since this has implications for both IDT performance and operation of SAW chemical sensors (3,7). The (semi)conducting layer (partly) shorts out the piezoelectric field, resulting in both velocity change and attenuation of the wave. These effects can be characterized in terms of the electromechanical (or piezoelectric) coupling constant, K^2. This is defined (3) as twice the magnitude of the short-circuit velocity change, that is,

$$K^2 = 2\left|\frac{\Delta v_0}{v_0}\right| = 2\frac{\Delta k}{k} \tag{3.7}$$

In physical terms, K^2 gives the fraction of the total potential energy of the wave that appears as ϕ(the potential generated at the surface by the SAW), the remainder being used to produce the elastic deformation of the substrate. As the potential wave passes through the substrate, it induces surface charges in the conducting layer. In electrical terms, we can think of this as a capacitative coupling between the substrate and the conducting layer. As K^2 increases, the amount of energy available for deformation is reduced, lowering the effective stiffness of the material and causing a slowing of the wave (Fig. 3.3).

If the film is not perfectly conducting, the coupling is no longer purely capacitative, but has an additional resistive component. In other words, the redistribution of the surface charges in response to ϕ does not occur instantaneously. As explained by Alder (32), the result is that the induced charge wave (φ_s) lags behind ϕ and never reaches its normal amplitude, so a com-

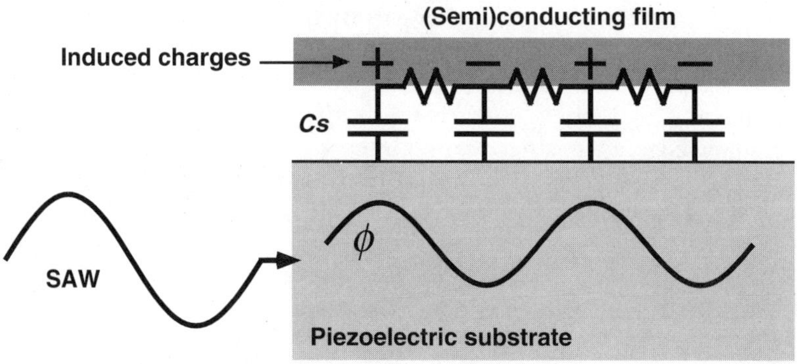

Figure 3.3. Model for the effect of a surface (semi)conducting film, showing distributed capacitance and film resistance toegther with the potential wave (ϕ) accompanying the SAW.

plete short circuit of the SAW is not obtained. Following the analysis of Datta (3), we can rewrite Equation 3.7 as

$$\frac{\Delta k}{k} = \frac{K^2}{2} = \frac{Z_0 v_0 W}{2}\left\{\frac{-\rho_{sc}}{\phi}\right\} \tag{3.8}$$

where Z_0 is the characteristic impedance of the device, W is the beam width (Fig. 3.1), and ρ_{sc} is the surface charge density. It is possible that the electrical properties of the surface film are such that (ρ_{sc}/ϕ) is complex, in which case Δk will also have real and imaginary components:

$$\Delta k = \beta + j\alpha \tag{3.9}$$

This in turn implies that the electric fields vary as a complex exponential function so that both attenuation (α is negative) and amplification (α is positive) are possible. It can be shown that for a thin surface layer of arbitrary conductance, the fractional velocity change and attenuation per wave number are

$$\frac{\Delta v}{v_0} = \frac{-\Delta\beta}{k} = -\frac{K^2}{2}\frac{1}{1 + (v_0 C_s/\sigma_s)^2} \tag{3.10}$$

$$\frac{\alpha}{k} = \frac{K^2}{2}\frac{v_0 C_s/\sigma_s}{1 + (v_0 C_s/\sigma_s)^2} \tag{3.11}$$

where C_s is the surface capacitance per unit length and σ_s is the sheet conductivity. When $v_0 C_s = \sigma_s$, $\alpha/k = -\beta/k \approx K^2/4$; for highly conducting films $-\beta/k \to K^2/2$, and for insulating films $-\beta/k \to 0$, while $\alpha/k \to 0$ in both cases (Fig. 3.4).

It remains then to explain how conditions can be created such that acoustic amplification is possible. Alder's analogy is useful here (32): consider the potential wave as being represented by a charged comb passing across the substrate surface producing an image charge in the surface film. If the film is semiconducting and has a current flowing through it, the charge carriers will then have a certain drift velocity v_s. If the comb is travelling at the same velocity, it will appear stationary relative to the charge carriers. The image charge in the surface film will therefore have an infinite amount of time to redistribute, so that the semiconducting film will appear as a perfect short circuit to the acoustic wave. If the comb travels faster than v_s, φ_s lags behind ϕ; if the comb travels slower, φ_s leads ϕ. By analogy with a simple a.c.-driven RC series circuit, it can be demonstrated that the former condition results in attenuation ($\alpha < 0$) while the latter results in amplification ($\alpha > 0$). In practical acoustic amplifiers, there is a slight air gap between the piezoelectric substrate and the semiconducting film in order to allow efficient interaction

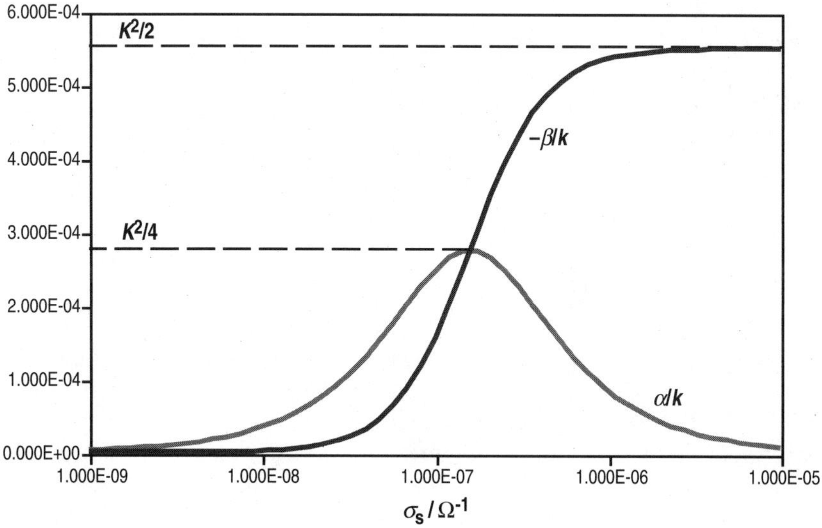

Figure 3.4. Plot of the attenuation and fractional velocity change as a function of surface sheet conductivity according to Equations 3.10 and 3.11. Calculation for $C_s = 0.5$ pF cm^{-1}, $K^2 = 0.0011$, and $\lambda = 60$ μm.

of ϕ with the film without interfering with the mechanical motion of the substrate surface (3,7,32).

The SAW chemical sensors employing (semi)conducting films have the film deposited directly on the substrate surface. Thus, the sensor response has contributions from both mass, elastic modulus, and conductivity effects. This has been clearly demonstrated by Roberts et al. (30) using monolayers of tetracyanoquinodimethane deposited by the Langmuir–Blodgett technique on a 98.6-MHz ST-X quartz SAW delay line. Other workers to exploit film conductivity for chemical sensing include Ricco et al. (6,33,34), Snow et al. (26,35), Falconer, Vetelino, and co-workers (36,37), and Barendsz et al. (38,39). D'Amico et al. (40–42) have employed metal films on SAW devices, but have not examined the contribution of film conductivity changes to observed sensor response. A similar comment applies to the metal phthalocyanine films employed by Rapp et al. (43).

3.6. OPERATION IN LIQUID ENVIRONMENTS

The early work on SAW chemical sensors was based entirely on gas–phase systems. The obvious question to ask, then, was whether these devices could also be operated in liquids, since the majority of potential chemical and biological sensor applications involve liquid–phase measurements. The answer is that this depends entirely on the mode of propagation and device format employed. As will be seen below, Rayleigh waves are rapidly attenuated by a semiinfinite liquid layer, making it impossible to operate conventional SAW devices in the liquid phase. There are many other acoustic modes and device formats available, however, that allow the use of surface-launched waves with liquid samples.

3.6.1. Thin Liquid Overlayer

The case of a thin liquid layer on the surface is identical to that for pressure loading. As mentioned previously, Rayleigh waves have both longitudinal and transverse motional components and can couple into a contacting fluid (7,44–46). The longitudinal component couples through the liquid viscosity while the transverse component couples through an acoustic mismatch (partial transmission due to a difference in acoustic impedance between the substrate and the liquid), this new mode of propagation being known as a "leaky" SAW wave (47). The relative contributions of these two mechanisms depend on the liquid viscosity and the thickness of the liquid layer, so that if the layer is very thin the attenuation can be quite moderate and a Rayleigh wave can still propagate. For fluid layers thicker than a few acoustic wavelengths, however, the losses are large and can result in total attenuation of the Rayleigh wave (48).

These effects can be seen clearly in the adsorption isotherms reported by Cheeke et al. (49) for superfluid helium films on a 143-MHz $LiNbO_3$ SAW device. The isotherms show a steady increase in Rayleigh wave attenuation up to the saturation vapor pressure, a maximum being observed in the curve due to the onset of superfluidity. At partial pressures close to p_0 (the saturation vapor pressure), however, there is a very rapid increase in attenuation due to multilayer buildup, that is, the condition $h \ll \lambda$ is no longer satisfied. In this particular case, only attenuation of the wave was observed, since the associated velocity change was too small to be measured. Velocity changes as reflected by Δf have been found, however, for nitrogen adsorption isotherms on thin sol-gel films using 97 MHz ST-X quartz delay lines (50–53). Under the conditions used, the effect of the adsorbed species can be described using Equation 3.6 in a more useful form:

$$\Delta f/f_0 = \kappa\ \Delta v/v_0 = -\kappa C_m f_0 m\delta \tag{3.12}$$

where κ is the fraction of the active surface area used, C_m is the mass sensitivity of the device ($C_m = k_1 + k_2$), δ is the number of adsorbed molecules per unit area, and m is the mass of one molecule of the adsorbing species. The same authors also used Equation 3.12 with adsorption isotherms for argon, carbon tetrachloride, toluene, cyclohexane, and acetone to measure pore-size distributions (54). A similar result has also been obtained for a ZnO-on-Si SAW resonator (55), although in this case Δf is given as varying linearly with f_0 rather than in a quadratic manner (presumably due to the porous nature of the ZnO film).

3.6.2. Thick Liquid Overlayer

When the liquid film thickness on a SAW device becomes appreciable or greater than the Rayleigh wavelength, attenuation of the wave is dramatically increased such that Rayleigh waves will cease to propagate across the device. These conditions occur for either multilayer buildup (e.g., BDDT type II isotherms) or immersion of the acoustically active side of the device in a liquid (semiinfinite layer). It is still possible, however, to operate a surface-acoustic wave device under these conditions by exploiting other acoustic modes (48,56).

For Lamb wave (FPW) devices, the effect of liquid loading can be modelled using equation 3.4 (Section 3.2.2) (16). As with the Rayleigh wave, the A_0 (lowest antisymmetrical) mode Lamb wave has both horizontal and vertical displacement components. The vertical component produces an evanescent pressure wave which is sensitive to liquid mass loading, giving rise to a frequency perturbation. The horizontal component couples viscously to generate a shear wave in the liquid, resulting in signal attenuation. By combining simultaneous measurements of attenuation and frequency of maximum transducer efficiency using a frequency pulse technique, it is therefore possible to simultaneously measure both liquid density and viscosity (57,58). White and co-workers report that attenuation varies linearly with $(\rho_f \eta_f)^{1/2}$ where ρ_f and η_f are the liquid density and shear viscosity, respectively. Taking into account viscous mass loading, the frequency of maximum transducer coupling (f_{MC}) was shown to be

$$f_{MC} = \frac{1}{\lambda}\left[\frac{B}{M + \rho_f K/k + (\rho_f \eta_f / 2\omega)^{1/2}}\right]^{1/2} \tag{3.13}$$

where $\omega = 2\pi f$, and B, M, and K are as defined previously (Equation 3.4, Section 3.2.2).

Shear-horizontal acoustic plate-mode devices also operate in liquid environments (59–62). Unlike Rayleigh and Lamb waves, these have no vertical displacement component, while the horizontal component is parallel to the device surface and perpendicular to the direction of propagation. This results in a device that shows comparatively little acoustic coupling into the liquid. The device is, however, sensitive to viscous coupling. Martin et al. (62) have shown that, for a simple Maxwellian fluid, the velocity and attenuation perturbations of the nth SH mode caused by an entrained liquid layer are

$$\alpha \approx \frac{c_v \eta_f}{2v_n} \mathbf{Re}\left[\frac{\gamma}{1 + j\omega\tau}\right] \tag{3.14}$$

$$\frac{\Delta v_n}{v_n} \approx -\frac{c_v \eta_f}{2\omega} \mathbf{Im}\left[\frac{\gamma}{1 + j\omega\tau}\right] \tag{3.15}$$

where

$$\gamma^2 = \left(\beta_n^2 - \frac{\omega^2 \rho_f}{\mu_f}\right) + j\frac{\omega \rho_f}{\eta_f}$$

$$\beta_n = \sqrt{(\omega/v_0)^2 - (n\pi/b)^2}$$

$$\tau = \eta_f/\mu_f$$

and b and μ_f are the membrane thickness and liquid shear modulus, respectively. This model provided a good fit to experimental measurements for glycerol/water mixtures, attenuation increasing and velocity change passing through a minimum with increasing $\sqrt{\eta_f}$.

3.7. VAPOR SORPTION AND TEMPERATURE EFFECTS WITH POLYMERIC FILMS

The majority of published descriptions of SAW chemical sensors for organic vapors exploit polymeric films of varying thicknesses to provide the required degree of sensitivity and selectivity. The detection process then involves two stages: buildup on the surface of the film (adsorption) followed by diffusion into the bulk of the film (absorption) (Fig. 3.5). These two processes will be collectively referred to here as "vapor sorption." Perturbation of f_0 or v_0 is then dependent on (a) the effect of the polymeric film in the absence of any

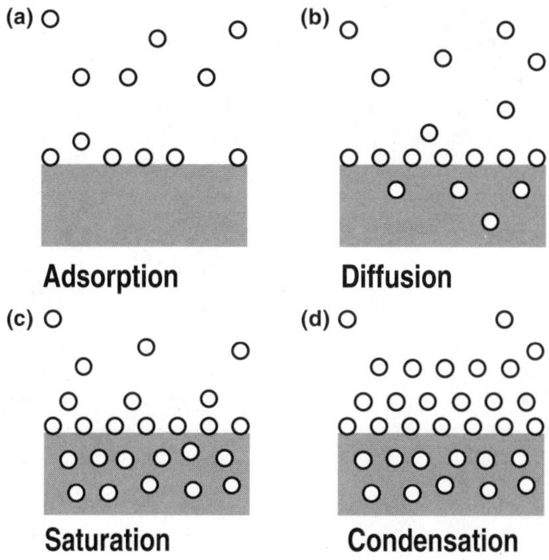

Figure 3.5. Stages in vapor sorption to a polymer film. (*a*) Initial rapid adsorption; (*b*) slower diffusion (absorption) into film with continual adsorption to resulting vacant sites; (*c*) saturation of the polymeric film with vapor molecules; and (*d*) multilayer buildup on the polymer surface.

sample vapor; (b) the effect of increasing mass due to vapor sorption; and (c) the effect of the sorbed vapor on polymer properties as they influence (a). In other words, the system can be used either for the detection of organic vapors *or* for the study of polymer properties. In general terms, we see from Section 3.4 that if the polymeric film is rigid or "stiff," we would expect its response to be that of a rigid, nonconducting film, that is, primarily due to increases in mass loading. On the other hand, more "liquid-like" or "mobile" polymers can be expected to show increasing film modulus-related effects. Thus, we are concerned not just with the amount of material present on the surface of a device, but also with its physical properties such as density, viscosity, and elasticity. On this basis, we can see that device temperature and operating frequency are also influential in determining sensor response.

Generally, the polymeric films applied to SAW devices are considered to be viscoelastic. That is, their properties are between those of a pure elastic solid (stress directly proportional to strain but independent of the rate of strain) and a pure viscous liquid (stress directly proportional to the rate of strain but independent of strain) (63). When such materials are subject to periodic stress, such as that caused by passage of an acoustic wave, the phase difference between the stress and the strain therefore lies between 0° and 90°, representing the extremes for purely elastic and viscous materials, respectively. The energy (in our case from the acoustic wave) is therefore partly stored (and recovered) by the material, and partly dissipated as heat. The degree to which energy is stored or lost (and thus the extent to which the acoustic wave is attenuated) depends on the ability of the polymeric chains and localized chain segments to redistribute within the bulk polymer. The former processes are slow, while the latter are fast. Clearly, polymer density and cross linking will have an affect here: the more closely packed the chains are, the greater the frictional forces between them and the harder it is for the polymer to "relax" from the applied deformations. In fact, "Every amorphous polymeric system has a glass transition temperature (T_g) below which the writhing thermal motions essentially cease" (63, page 3). The significance of this transition is discussd in greater detail below. For now, we simply note that when a vapor is sorbed by a polymer, it will have several effects: first, the total mass of the film is increased; second, there is an accompanying volume change (polymer swelling) as the sorbed vapor intercalates between the polymer chains; and third, there is a corresponding change in the viscoelastic properties of the polymer. Exactly how these effects combine, and what the net effect on SAW velocity and attenuation are for a given polymer/vapor pair has been the subject of some debate and detailed investigation recently within the chemical sensor literature. Since a correct understanding of these processes is essential to most SAW chemical sensor applications, the development of these theories is treated here in some detail (Table 3.5).

TABLE 3.5. Studies of Vapor Sorption into Polymer Films

Device	f_0 (MHz)	Polymer	h (μm)	Vapor	Shift	Ref.
SAW	97	Polyimide	1.8	Methanol, N_2O	+	50,54,64
		Polystyrene	0.20	Methanol, toluene	−	64
		Polysiloxane	0.36	Methanol	−	64
SAW	52	Polyimide	3.0	Methanol	−	65
		Polycarbonate	≤0.58	Toluene	+, −	65
SAW	97	SBS[a]	0.34	*n*-Pentane	+, −	66
				CH_2Cl_2	+, −	66
				CCl_3CH_3	+, −	66
				CH_2Br_2	−	66

[a] SBS = styrene–butadiene–styrene copolymer.

3.7.1. Vapor Sorption into a Polymeric Film

Historically, Martin et al. (50) made the assumption that vapor sorption at low gas-phase concentrations into polymers well below their glass transition temperature (T_g) would result in film "stiffening," that is, increasing shear modulus, leading to a positive frequency shift, as indicated by Equation 3.5. This effect can be understood by considering the effect of material stiffness on acoustic wave velocity: v increases with increasing stiffness. If the polymer film is rigidly bound to the substrate surface, vapor-induced swelling will initially cause a stiffening of the (constrained) interfacial region (Fig. 3.6). At higher gas-phase concentrations, however, "softening" will occur due to solvent plasticization of the bulk polymer film. In the former case, Martin et al. showed that the cumulative frequency response to vapor diffusion as a function of time into a thin polymer film could be described by:

$$\Delta f(t) = \Delta f_{\max}\left(1 - \frac{2}{\pi^2}\sum_{n=1}^{\infty}\frac{\exp\{-[\pi(n-\frac{1}{2})/h]^2 Dt\}}{(n-\frac{1}{2})^2}\right) \tag{3.16}$$

where h is the film thickness and D is the diffusion coefficient. It is perhaps helpful to reiterate at this point the underlying assumptions of Equation 3.16 (54): (1) one-dimensional Fickian diffusion occurs; (2) the diffusion coefficient may be assumed to be constant; and (3) SAW frequency response is linear with respect to the molar concentration of vapor within the film. Experimental agreement with Equation 3.16 was reported for 5% nitrous oxide and a partial pressure (p/p_0) of methanol of 0.05 diffusing into a thin polyimide film (50). Positive frequency shifts were reported for increas-

(a)

(b)

(c)

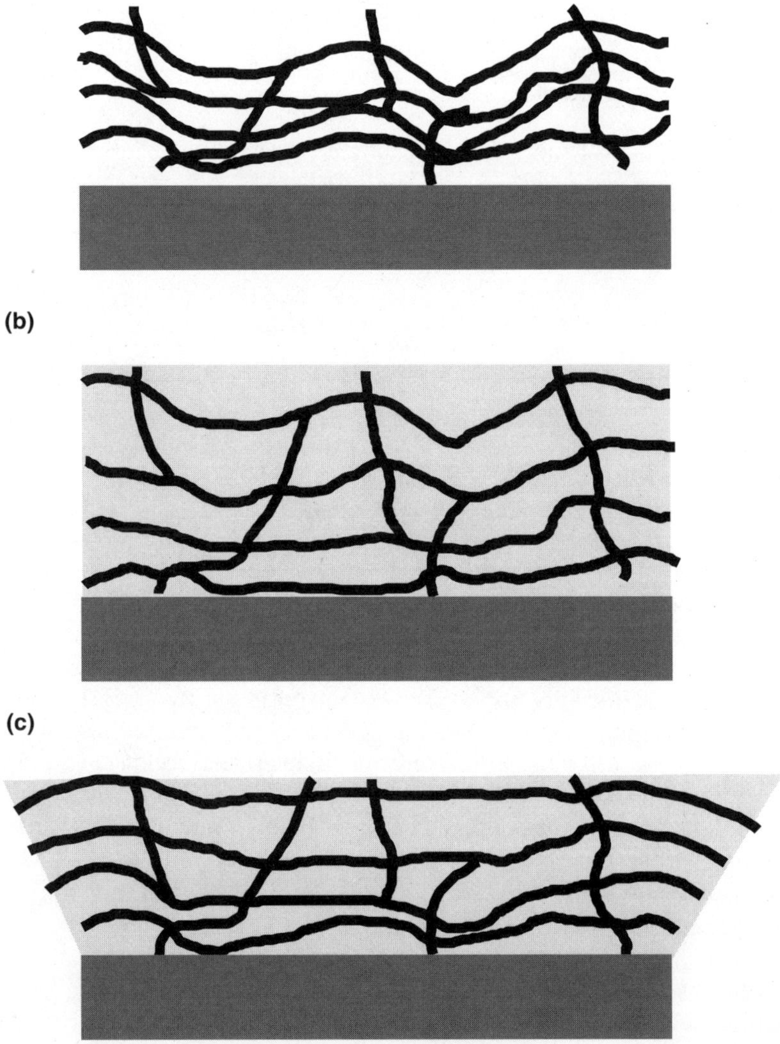

Figure 3.6. Models for polymer–vapor interaction effects. (*a*) Polymer in absence of vapor. (*b*) Polymer swollen by vapor but unconstrained by the substrate surface. (*c*) Polymer swollen by vapor *and* physically constrained by the substrate surface.

ing methanol concentration over the range $p/p_0 = 0.005 - 0.061$ with a 1.8-μm thick polyimide film while D was found to vary over the range $0.2–0.8 \times 10^{-10}$ cm^2 s^{-1} and to vary linearly with methanol concentration (54). The same authors also reported results for polyimide, polystyrene, and polysiloxane films using nitrous oxide, methanol, ethanol, *n*-propanol, toluene, and isooctane (64). Negative frequency shifts were reported for toluene, while results for methanol over the extended concentration range $p/p_0 = 0.01-0.8$ showed a maximum in the (positive) frequency shift at $p/p_0 \approx 0.25$. The positive frequency shifts observed with methanol were therefore attributed to the stiffening/plasticizing effects of this species on polyimide, although it was also recognized that polymer swelling (resulting in increasing h) would also contribute a positive frequency shift (54).

The issue of possible film elastic-tightening effects was futher addressed by Bartley and Dominguez (65). Again it was assumed that the film is rigidly attatched to the substrate surface. Under these conditions, polymer swelling can be described in terms of compression of an identical, unattached film to the same surface dimensions. Bartley and Dominguez developed equations for the effect of such a film on Rayleigh wave propagation, and presented some experimental findings. They reported negative frequency shifts for an equivalent concentration of methanol (sample vapors were generated at 0°C as opposed to 25°C in the work of Martin et al.) and a 3.0-μm thick polyimide film. The diffusion coefficients obtained were $\approx 0.2 \times 10^{-10}$ cm^2 s^{-1}. Extrapolating the results of Martin et al. gave a value an order of magnitude greater at the same partial pressure. However, this may not be significant because the partial pressure value is well outside the calibration range of Martin et al., while the frequency response is nonlinear (50,54,64). This not withstanding, it remains that two experiments similar in operating conditions (with some notable exceptions) give completely opposite frequency shifts for methanol sorption into polyimide films. Bartley and Dominguez also reported both positive and negative frequency shifts for toluene sorption into polycarbonate, a combination that the authors expected to be most sensitive toward compressive tension effects relative to a pure mass-based response.

Martin and Frye subsequently readdressed the issue of film property changes on vapor sorption using a parametric approach (66). This takes the Tiersten equation (from which Equation 3.5 is derived) and expresses the effect of elastic modulus and mass loading effects in terms of the fractional velocity shift and attenuation per wave number:

$$\Delta v/v_0 = -c_1\omega\Delta(h\rho') - c_2\omega\Delta\left[\frac{h\mu(\omega\tau)^2}{1 + (\omega\tau)^2}\right] \tag{3.17}$$

$$\Delta\alpha/k = c_2\omega\Delta\left(\frac{h\mu(\omega\tau)^2}{1 + (\omega\tau)^2}\right) \tag{3.18}$$

where τ is the relaxation time of the polymer and c_1 and c_2 are material constants analagous to k_1 and k_2. Martin and Frye plotted $\Delta\alpha/k$ vs. $\Delta v/v_0$ for sorption of vapors into a polystyrene–polybutadiene triblock copolymer over the partial pressure range 0–80% (at 10°C). Both positive and negative frequency shifts were observed for pentane and a series of halogenated hydrocarbons. Given the relatively high vapor concentrations, this was interpreted as due to increasing plasticizing effects offsetting the negative shift due to pure mass loading. It was found that the observed results could be adequately explained in terms of a simple Maxwellian viscoelastic model in which shear viscosity (η) represents the interchain frictional resistance while the shear stiffness (μ) represents entropy-related restoring forces—it is expected that vapor sorption will decrease η but leave μ relatively unchanged.

3.7.2. Temperature Effects on Polymeric Films

So far we have considered only the effect of vapor sorption by the film. Temperature can also have a considerable influence on film viscosity, stiffness, and volume, however. Before continuing with the analysis of vapor sorption effects, it is therefore necessary to examine the role of temperature. As the temperature of an amorphous polymer is increased, the thermal motion of individual chains and chain segments also increases and the polymer swells. At a certain temperature, sufficient thermal energy is available for significant long-range readjustments of chain confirmations—the polymer ceases to show "glassy" characteristics, and becomes more "rubbery" (63). In addition, there can be marked changes in specific heats and thermal expansion rates. The temperature at which the former effects occur is highly frequency dependent, while the latter is found to be frequency independent. Following the convention of Grate et al. (67), we therefore denote these transition temperatures as the dynamic (T_α) and static (T_g) glass transition temperatures, respectively. (We discuss the practical measurement of these transitions in Chapter 6.)

In the third of their original papers on SAW sensors, Wohltjen and Dessy (68) first proposed that such devices could be used as probes of thin polymer film properties, specifically the glass transition temperature. The majority of their experiments were performed on bulk polymer samples mechanically clamped to the surface of the SAW device, and so are not of immediate relevance here. They did, however, observe T_α = 150°C for a thin poly(methyl methacrylate) film on a 31-MHz SAW device, compared to 100°C for conventional low-frequency measurements. Amongst the work that followed, key observations were (1) that there was a continual change in frequency for polymer-coated SAWs as the temperature was ramped, even after correcting for the temperature-dependence of the uncoated device (69) and (2) that localized minima in the frequency–temperature data could be observed for specific polymers at temperatures corresponding to either the melting point or dynamic glass transition (70). Ballantine and Wohltjen proposed that the former effects could be explained by polymer swelling and flow under the conditions of incomplete surface coverage or to changes in μ' and λ' accompanying T_g or T_α (Equation 3.5). Under conditions of complete surface coverage, it was assumed that polymer swelling alone would not cause a frequency shift, since the first term in Equation 3.5 can be identified as a dependence on mass per unit area—in other words, decreasing density would be compensated for by increasing film thickness over the same surface area. Martin and Frye (66) also observed a transition for a polystyrene/polybutadiene copolymer on a 97-MHz SAW device using the parametric approach described above. Similarly, Grate et al. (67) measured both frequency and amplitude changes with changing temperature for a variety of pure polymeric films using a 5-MHz flexural plate wave device. Both T_g and T_α transitions were observed.

Since Grate et al. have already made a highly detailed analysis of the foregoing results (67), we will not reproduce it here other than to summarize some of the key conclusions. In essence these are

1. SAW and FPW devices may be used to observe both static and dynamic transitions in thin polymer films.
2. Both velocity (frequency) and attenuation measurements should be made to best observe such transitions.
3. Different results for film vapor sorption may be obtained depending on sensor temperature relative to T_g (especially when sensor response is diffusion controlled) due to differences in polymer free volume.
4. Sensor temperature can have complex consequences for response depending on the value of T_g and the extent to which operating frequency raises T_α.

Points 3 and 4 bring us back to the issue of sensor response to vapor sorption by a thin, isotropic polymer film. When a film is held above its T_g, thermally induced swelling will allow faster diffusion and a greater mass uptake of vapor. This in turn will cause further swelling, which may alter not just the film modulus, but also T_α. In other words, it is quite possible that, as vapor concentration increases, nonlinear calibration curves, positive frequency shifts, and unusual attenuation may be observed. This also has implications for the sorption–desorption profiles usually recorded in SAW chemical sensor measurements (Fig. 3.7), since these now reflect the rate and extent of vapor adsorption (surface phenomena), vapor absorption (bulk diffusion

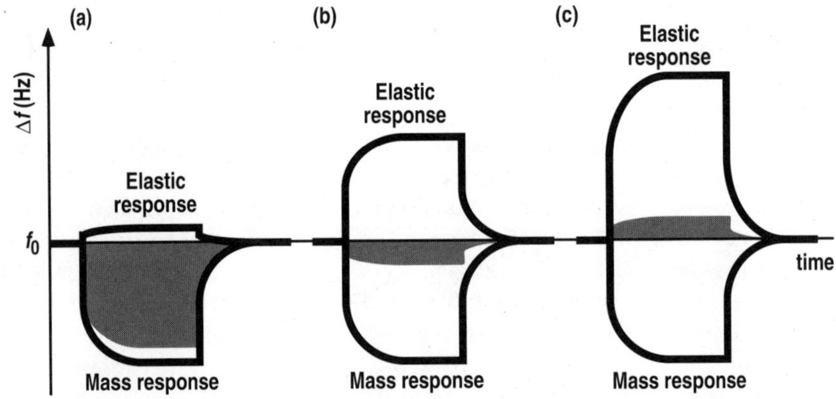

Figure 3.7. Effect of added mass and polymer property changes on SAW sensor response to vapor sorption. (*a*) $T_g, T_\alpha > T_{\text{sensor}}$ and elastic effects are negligible. Sensor response is given by Equation 3.6. (*b*) $T_g < T_{\text{sensor}} < T_\alpha$. Significant elastic effects. (*c*) As *b*, but elastic effects exceed change in mass and give a positive frequency response.

phenomena), and film swelling (modulus effects). Both Grate et al. (71) and Martin et al. (72) have proposed models to explain the combined effects of temperature and vapor sorption on sensor response.

3.7.3. Combined Temperature and Vapor Sorption

Grate et al. (71) report an exponential dependence of frequency response on temperature for a variety of polymers on sorption of isooctane, 1,2-dichloroethane and toluene over the range 25–45°C. This response consists of contributions from both thermal expansion and vapor sorption by the polymer film, the latter decreasing rapidly with increasing temperature as might be expected on kinetic grounds. The experimental results confirm that the SAW responses are systematically greater in magnitude than expected on a pure mass basis, which they explain by a decrease in polymer modulus due to vapor-induced swelling. Using partition coefficient measurements, they were able to calculate the "excess response" of modulus effects over mass effects. Further, by relating the coefficient of volume expansion due to vapor sorption to the thermal expansion coefficient, an equation could be derived that allows prediction of the combined effects on SAW sensor response:

$$\Delta f_v = (\Delta f_s C_v K/\rho_s) + (\Delta f_s C_v K/\rho_L) A_T \tag{3.19}$$

where Δf_v is the sensor response, Δf_s is the shift produced by the polymeric coating alone, K is the partition coefficient, ρ_s and ρ_L are the film and liquid densities, respectively, and A_T is a scaling factor related to the film thermal expansion coefficient. Once again, the detailed analysis provided by these authors will not be reproduced, but some of their conclusions should be stated:

1. Although vapor sorption can decrease the polymer modulus, giving a larger excess response, a point is reached where further vapor concentration increases do not produce significant changes in the modulus, resulting in nonlinear calibration curves. This in turn implies that sensor calibration curves may not always be equated with sorption isotherms.
2. Swelling due to vapor sorption may be estimated from partition coefficients and liquid densities. These volume changes are significantly greater than those from pure thermal expansion over small temperature ranges.
3. The derived model estimates that swelling effects will be four to six times greater than mass effects for polymers and liquids having similar densities.

Although the preceding model generally provided an explanation for vapor responses that showed nonlinear calibrations or magnitudes greater than predicted on a mass basis, it did not explain the positive frequency shifts observed by Martin and Frye (66). These authors subsequently explained this discrepancy by reconsidering the boundary conditions used in deriving Equation 3.5 (72). First, viscoelastic materials both store and dissipate acoustic energy, resulting in SAW attenuation, as noted above. This effect is not accounted for in the Tiersten–Wohltjen model. Second, for Equation 3.5 to be valid, the condition $h \ll \lambda$ must be satisfied. As can be seen from Table 3.5, values of h used in the studies reported here have ranged from 5% of λ downward; in fact, the condition $h \leq 0.2\% \, \lambda$ has been suggested. Clearly both temperature and vapor sorption will increase h: Grate et al. (71) have calculated that this change is typically no more than about 1%, which would not appear to present a problem. The analysis of Martin et al. (72), however, clearly shows that a film satisfying the condition $h \ll \lambda$ may still *appear* acoustically "thick" when it is significantly softer than the substrate. This is because the displacement variations in the film will then be much smaller than λ, breaking the implicit assumption in the model of synchronous film motion.

The shear modulus (G) of a linear, isotropic polymer is a complex quantity comprising a storage (real) and loss (imaginary) component, where G'' is the stress component 90° out of phase with the strain. A similar expression is obtained for the bulk modulus K:

$$G = G' + jG'' \tag{3.20}$$

$$K = K' + jK'' \tag{3.21}$$

In terms of the chain segment relaxation time (τ) and the angular frequency (ω), the glassy and rubbery regimes of polymer behavior correspond to the conditions $\omega\tau \gg 1$ and $\omega\tau \ll 1$, respectively, while for $\omega\tau \approx 1$, G'' shows a maximum. Considering the region of the film in immediate contact with the substrate during propagation of a Rayleigh wave, since the film is assumed rigidly attatched to the substrate, the particle displacements (in-plane) within the film track the deformation of the substrate surface. As this displacement propagates deeper into the bulk of a thick film, however, viscoelastic properties result in an inertial lag of the particle displacements (cross-film), leading to a shear film deformation. Martin et al. represent this effect in terms of a complex wave propagation factor (γ) such that the normalized perturbation due to the film is

$$\frac{\Delta\gamma}{k_0} = \frac{\Delta\alpha}{k_0} - j\frac{\Delta v}{v_0} \tag{3.22}$$

where k, α, and v are as defined previously. It was shown that the ratio of film strains generated by cross-film to in-plane displacement gradients was

$$R = \frac{Afv_0\rho h}{|G|} \tag{3.23}$$

where A is a material constant ($\approx$1.9 for ST-X quartz). Equation 3.23 allows the identification of acoustically thick and thin films: for $R \ll 1$, in-plane displacement gradients dominate and the film is acoustically thin; for $R \geq 1$, cross-film gradients dominate and the film is acoustically thick.

Following this approach, Martin et al. then developed models for SAW response from acoustically thin and thick films. In the first case (model 1), the model reduces to Equation 3.5 for acoustically thin, *elastic* films ($R \ll 1$, $G'' = 0$). For acoustically thin *viscoelastic* films, however, the model yields the result

$$\frac{\Delta v}{v_0} \approx -\omega h\left[(c_1 + c_2 + c_3)\rho' - \frac{(c_1 + 4c_3)}{v_0^2}G'\right] \tag{3.24}$$

$$\frac{\Delta\alpha}{k_0} = \frac{\omega h}{v_0^2}(c_1 + 4c_3)G'' \tag{3.25}$$

TABLE 3.6. Definition of Acoustically "Thin" and "Thick" Films for a 100-MHz ST-X Quartz SAW Device Using Equation 3.23 (72)

h (μm)	h/λ (%)	G'	R	State
0.1	0.30	10^{10}	$\ll 1$	Thin
		10^{8}	≥ 1	Thick
<0.020	0.06	10^{7}	$\ll 1$	Thin
		10^{11}	$\ll 1$	Thin
≥ 2	6.33	10^{7}	≥ 1	Thick
		10^{11}	≥ 1	Thick

where for ST-X quartz c_1, c_2, and c_3 are 1.3×10^{-9}, 1.421×10^{-7}, and 0.615×10^{-7} cm^2 s g^{-1}, respectively. The situation for acoustically thick films (model 2) is understandably more complex, and the resulting equations will not be reproduced here. It should be noted, however, that for appropriate values of R and G', model 2 reduces to Equations 3.24 and 3.25 as expected.

Particle displacements within a thick film were found to arise from the superposition of two shear waves and one compressional wave generated at the substrate–film interface by the SAW. These waves reflect back from the upper surface of the film and interfere constructively or destructively, depending on their phase shifts. For typical polymer films, the compressional wave simply results in a vertical translation of the film. The shear waves, however, influence both attenuation and velocity of the propagating SAW. Of particular importance is the prediction that the shear wave generated by particle displacement polarized in the direction of SAW propagation (which dominates for ST-X quartz SAW devices) can experience a film resonance condition such that, in the case of an *elastic* acoustically thick film, large positive or negative velocity perturbations can occur while the mechanical surface acoustic impedance becomes infinite. For *viscoelastic* acoustically thick films, Δv and $\Delta \alpha$ both remain finite, but can be large. To put this into perspective, Table 3.6 summarizes the calculations of Martin et al. as to what conditions represent acoustically thin and thick films. These authors have also provided a theoretical method for predicting the storage and loss components of G and K for any apparent probe frequency where the latter is the actual transition frequency perturbed by temperature and/or vapor sorption.

3.7.4. Summary

As a polymer passes through its T_α, there are corresponding decreases in G' and K', and maxima in G'' and K''. These were originally assumed to yield corresponding velocity and attenuation changes directly (66). Although such transitions have been observed (as described in the references cited above), the temperatures and frequencies at which they were observed have not always matched those expected from other measurements (71). This problem may be explained using the film resonance model (72), which shows that the assumption above is valid only for films that remain acoustically thin across the whole modulus range (from glassy to rubbery) due to the rapid decrease in G' accompanying T_α. Experimental results obtained for poly(isobutylene) films of various thicknesses on a 97-MHz ST-X SAW device show that for acoustically thick films, dynamic glass transitions cannot be observed if resonance effects occur at temperatures near T_α, and that observed attenuation maxima are subject to thickness-dependent film resonance effects. These problems may be mitigated by employing flexural plate wave (FPW) devices rather than SAW devices, since the former can provide similar sensitivities at lower frequencies, allowing measurements to be performed entirely within the acoustically thin regime.

In conclusion, then, we see that the effects of temperature and vapor sorption on polymer-coated SAW response are complex phenomena in which film properties, thickness, temperature, and measurement frequency are critical parameters. When $h \ll \lambda$ *and* the film is acoustically thin under all conditions (Equation 3.23), Equation 3.5 may be used to predict the frequency response due to vapor sorption in which the film density and modulii are modified by the vapor; the response thus consists of both mass and elastic modulus effects, and calibrations may be nonlinear, even for systems having linear adsorption isotherms. A more accurate interpretation of sensor response data for acoustically thin films can be obtained using Equations 3.24 and 3.25, however, which removes some of the assumptions implicit in Equation 3.5. Under conditions for which a film is acoustically thick, film resonance effects may result in anomalous data, including positive frequency shifts and anomalous T_α measurements. In either case, simultaneous measurements of both SAW velocity and attenuation changes allow a much better understanding of polymer properties with respect to both vapor sorption and temperature efffects to be achieved. Finally, in circumstances where film modulii are not known, temperature and vapor sorption effects may be estimated using either the polymer-swelling model of Grate et al. (71) or the modulii "master curves" of Martin et al. (72).

3.8. DIELECTRIC AND SOLUTION CONDUCTIVITY EFFECTS

Since surface conductivity has already been shown to be an important mechanism of SAW velocity perturbation, it is not surprising that this has been exploited for liquid-phase detection of ionic species using both SH-APM and

Love wave devices. The influence of the dielectric constant of the medium directly in contact with the substrate, while recognized as a possible factor [33,39,64], has until recently been considered negligible. Recent reports have demonstrated that dielectric effects can, however, be significant [46,73]. There are two possible mechanisms for dielectric effects: dielectric loss and interdigital capacitance changes. Here we discuss the former in conjunction with solution conductivity effects; the latter mechanism is discussed in Chapter 4.

3.8.1. Dielectric Loss Effects

The dielectric constant of a particular substance is related to its molecular dipole moment and polarizability (ease of dipole induction by an external applied electric field) through the Debye equation:

$$N_v\left[\frac{(\chi + \mu_D^2)}{3\varepsilon_0 kT}\right] = \frac{3(\varepsilon_r - 1)}{(\varepsilon_r + 2)} \tag{3.26}$$

where N_v is the number of molecules per unit volume, χ is the polarizability, μ_D is the dipole moment, ε_0 is the permittivity of free space, k is Boltzmann's constant, T is temperature, and ε_r is the material relative permittivity. When an electric field is applied to a dielectric medium, polarization occurs in which dipoles and induced dipoles become aligned with the field. There are three basic mechanisms for this: (1) electron displacement relative to atomic nucleii, (2) atomic displacement within a molecule, and (3) molecular rotation and translation (74). These processes can be measured in terms of the total, electronic, atomic, and orientation polarizibilities:

$$\chi_T = \chi_e + \chi_a + \chi_o \tag{3.27}$$

Electronic and atomic polarizations occur on timescales corresponding to frequencies in excess of 10^{12} Hz, and so have little or no effect at typical SAW operating frequencies. Molecular reorientations, however, occur on the picosecond timescale (10^{10}–10^{12} Hz), whereas polymer relaxation and electrode effects can be observed at frequencies from d.c. to 10 GHz. It is therefore possible that the electric potential wave accompanying the SAW can couple through dipole–induced dipole reorientation, resulting in a dielectric loss term. Such effects will be most pronounced for piezoelectric substrates having a high electromechanical coupling constant (K^2), such as lithium niobate and lithuium tantalate. Baer and Flory (46) report a dramatic increase in attenuation for their STW sensor for loading with propanol vs.

dodecane, even though these two solvents have the same viscosity–density product.

3.8.2. Solution Conductivity Effects

Various SH-APM and Love wave devices have been employed as sensors for ions in aqueous solution (75–79). In essence, the considerations here are identical to the case of a (semi)conducting layer, with the difference that the charge carriers are ions in solution and therefore will have lower drift velocities than electrons or holes in the former case. Niemczyk et al. (75) have modeled the interaction of ions in solution with the electric potential wave ϕ in terms of the dielectric constants of the substrate (ε_s) and solution (ε_f) and the liquid conductivity (σ_f). Following the work of Alder (32), the dielectric relaxation frequency (ω_R) is given by

$$\omega_R = \frac{\sigma}{\varepsilon_s + \varepsilon_f} \tag{3.28}$$

When the applied frequency is greater than ω_R, ions in solution cannot redistribute fast enough to follow the changing potential. The result is that the capacitive double layer is never fully formed, so the electric field penetrates into the bulk solution. The electrical interaction can then be reduced to an equivalent circuit composed of two capacitors and an impedance representing ε_s, ε_f, and σ_f, respectively. By solving for the power flow along the APM device using a perturbational analysis, Niemczyk et al. derived the following relationships:

$$\frac{\Delta v}{v_0} \approx -\frac{K^2}{2}\left(\frac{\varepsilon_s + \varepsilon_0}{\varepsilon_s + \varepsilon_f}\right)\frac{\sigma_f^2}{\sigma_f^2 + \omega^2(\varepsilon_s + \varepsilon_f)^2} \tag{3.29}$$

$$\frac{\alpha}{k} \approx \frac{K^2}{2}\left(\frac{\varepsilon_s + \varepsilon_0}{\varepsilon_s + \varepsilon_f}\right)\frac{\omega\sigma_f(\varepsilon_s + \varepsilon_f)}{\sigma_f^2 + \omega^2(\varepsilon_s + \varepsilon_f)^2} \tag{3.30}$$

Equations 3.29 and 3.30 provide a reasonable fit to the experimental data given by Niemczyk et al., although the model does not include either viscosity effects or the effects of mixed solvent systems and high salt concentrations on the effective value of ε_f. This acoustoelectric interaction has the least effect for low-conductivity solutions $[\sigma_f \leq \omega(\varepsilon_s + \varepsilon_f)]$; for higher conductivities, it should either be taken into account or the interaction "screened out" by deposition of a metal film onto the device surface.

REFERENCES

1. R. M. White and F. W. Voltmer, *Appl. Phys. Lett.* **1965,** *7,* 314.
2. B. A. Auld, *Acoustic Fields and Waves in Solids,* Vol. 2, Wiley Interscience, New York, 1973, pp. 88–94.
3. S. Datta, *Surface Acoustic Wave Devices,* Prentice-Hall, Englewood Cliffs, NJ, 1986.
4. V. M. Ristic, *Principles of Acoustic Devices,* Wiley, New York, 1983.
5. M. Feldmann and J. Hénaff, *Surface Acoustic Waves for Signal Processing,* Artech House, Norwood, MA, 1986.
6. A. J. Ricco, S. J. Martin, and T. E. Zipperian, *Sens. Actuators*, **1985,** *8,* 319.
7. B. A. Auld, *Acoustic Fields and Waves in Solids,* Vols. 1 and 2, Wiley Interscience, New York, 1973.
8. M. R. Daniel and J. de Klerk, *Appl. Phys. Lett.,* **1970,** *16,* 30.
9. A. J. Budreau and P. H. Carr, *Appl. Phys. Lett.,* **1971,** *18,* 239.
10. D. Hauden, M. Michel, G. Bardeche, and J.-J. Gagnepain, *Appl. Phys. Lett.,* **1977,** *31,* 315.
11. D. Hauden, G. Jaillet, and R. Coquerel, *IEEE Ultrason. Symp. (Proc.),* **1981,** 148.
12. J. Neumeister, R. Thum, and E. Lüder, *Sens. Actuators A* , **1990,** *A21-23,* 670.
13. M. Viens and J. D. N. Cheeke, *Sens. Actuators A* , **1990,** *A24,* 209.
14. A. J. Slobodnik, *J. Appl. Phys.,* **1972,** *43,* 2565.
15. M. R. Risch, *Sens. Actuators,* **1984,** *6,* 127.
16. S. W. Wenzel and R. M. White, *IEEE Ultrason. Symp. (Proc.),* **1988,** 563.
17. R. Inaba, Y. Kasahara, and K. Wasa, *IEEE Ultrason. Symp. (Proc.),* **1982,** 312.
18. S. G. Joshi, *IEEE Ultrason. Symp. (Proc.),* **1982,** 317.
19. E. Gatti, A. Palma, and E. Verona, *Sens. Actuators,* **1983,** 45.
20. P. V. H. Sabine, *Electron. Lett.,* **1970,** *6,* 149.
21. *Recent Developments in Surface Acoustic Waves,* D. F. Parker and G. A. Maugin, Eds., Springer-Verlag, New York, 1987.
22. A. A. Maradudin in *Recent Developments in Surface Acoustic Waves,* D. F. Parker and G. A. Maugin, Eds., Springer-Verlag, New York, 1987, pp. 100–128.
23. M. Thompson, D. C. Stone, and R. Nisman, *Anal. Chim. Acta,* **1991,** *248,* 143.
24. H. Wohltjen, *Sens. Actuators,* **1984,** *5,* 307.
25. H. F. Tiersten, *J. Appl. Phys.,* **1969,** *40,* 770.
26. A. W. Snow, W. R. Barger, M. Klusty, H. Wohltjen, and N. L. Jarvis, *Langmuir* **1986,** *2,* 513.
27. H. Wohltjen, D. S. Ballantine, and N. L. Jarvis, in *Chemical Sensors and Microinstrumentation,* R. W. Murray, R. E. Dessy, W. R. Heineman, J. Janata, and W. R. Seitz, Eds., *ACS Symp. Ser.* **1989,** *403,* 157.

28. H. Wohltjen, A. W. Snow, W. R. Barger, and D. S. Ballantine, *IEEE Trans. Ultrason. Ferroelec. Freq. Control,* **1987,** *UFFC-34,* 172.

29. J. W. Grate and M. Klusty, *Anal. Chem.,* **1991,** *63,* 1719.

30. G. G. Roberts, B. Holcroft, A. Barraud, and J. Richard, *Thin Solid Films,* **1988,** *160,* 53.

31. B. Holcroft and G. G. Roberts, *Thin Solid Films,* **1988,** *160,* 445.

32. R. Alder, *IEEE Trans. Sonics Ultrason.,* **1971,** *SU-18,* 115.

33. S. J. Martin, A. J. Ricco, D. S. Ginley, and T. E. Zipperian, *IEEE Trans. Ultrason. Ferroelec. Freq. Control,* **1987,** *UFFC-34,* 142.

34. A. J. Ricco and S. J. Martin, *Thin Solid Films,* **1991,** *206,* 94.

35. W. R. Barger, M. A. Klusty, A. W. Snow, J. W. Grate, D. S. Ballantine, and H. Wohltjen, in *Sensor Science and Technology,* B. Schumm, C. C. Liu, and R. A. Powers, Eds., *Electrochem. Soc. Proc.,* **1987,** *87-15,* 198.

36. R. Lec, J. F. Vetelino, R. S. Falconer, and Z. Xu, *IEEE Ultrason. Symp. (Proc.),* **1988,** 585.

37. R. S. Falconer, J. F. Vetelino, D. J. Smith, and M. J. Osborn, *IEEE Ultrason. Symp. (Proc.),* **1990,** 315.

38. A. W. Barendsz, J. C. Vis, M. S. Nieuwenhuizen, E. Nieuwkoop, M. J. Vellekoop, W. J. Ghijsen, and A. Venema, *IEEE Ultrason. Symp. (Proc.),* **1985,** 586.

39. M. S. Nieuwenhuizen, A. W. Barendsz, E. Nieuwkoop, M. J. Vellekoop, and A. Venema, *Electron. Lett.* **1986,** *22,* 184.

40. A. D'Amico, A. Palma, and E. Verona, *Appl. Phys. Lett.,* **1982,** *41,* 300.

41. A. D'Amico, A. Palma, and E. Verona, *Sens. Actuators,* **1982/83,** *3,* 31.

42. A. D'Amico, A. Petri, P. Verardi, and E. Verona, *IEEE Ultrason. Symp. (Proc.),* **1987,** 633.

43. M. Rapp, D. Binz, I. Kabbe, M. Von Schickus, S. Hunklinger, H. Fuchs, W. Schrepp, and B. Fleischmann, *Sens. Actuators B,* **1991,** *B4,* 103.

44. H. L. Bertoni in *Rayleigh-Wave Theory and Application,* E. A. Ash and E. G. S. Paige, Eds., *Springer Series on Wave Phenomena,* Vol. 2, Springer-Verlag, New York, 1985, pp. 274–290.

45. A. J. Slobodnik in *Acoustic Surface Waves,* A. A. Oliner, Ed., *Topics in Applied Physics,* Vol. 24, Springer-Verlag, New York, 1978, pp. 225–303.

46. R. L. Baer and C. A. Flory, *IEEE Ultrason. Symp. (Proc.),* **1991,** *1,* 279.

47. J. D. N. Cheeke and P. Morisseau, *J. Low Temp. Phys.* **1982,** *46,* 319.

48. G. S. Calabrese, H. Wohltjen, and M. K. Roy, *Anal. Chem.* **1987,** *59,* 833.

49. J. D. N. Cheeke, P. Morisseau, and M. Poirier, *Phys. Lett. A* **1982,** *88A* , 359.

50. S. J. Martin, G. C. Frye, A. J. Ricco, and T. E. Zipperian, *IEEE Ultrason. Symp. (Proc.),* **1987,** *1,* 563.

51. A. J. Ricco, G. C. Frye, and S. J. Martin, *Langmuir,* **1989,** *5,* 273.

52. C. L. Glaves, G. C. Frye, D. M. Smith, C. J. Brinker, A. Datye, A. J. Ricco, and S. J. Martin, *Langmuir,* **1989,** *5,* 459.

53. G. C. Frye, A. J. Ricco, S. J. Martin, and C. J. Brinker in *Better Ceramics Through Chemistry III,* C. J. Brinker, D. E. Clark, and D. R. Ulrich, Eds., Materials Research Society Symposium Proceeding, Vol. 121, April 5-8 Reno Nevada, 1988, p. 349.
54. G. C. Frye, S. J. Martin, A. J. Ricco, and C. J. Brinker in *Chemical Sensors and Microinstrumentation,* R. W. Murray, R. E. Dessy, W. R. Heineman, J. Janata, and W. R. Seitz, Eds., ACS Symposium Series, Vol. 403, American Chemical Society, Washington, DC, 1989, p. 208.
55. S. J. Martin, K. S. Schweizer, A. J. Ricco, and T. E. Zipperian, *Proceedings of the International Conference on Solid-State Sensors and Actuators,* Pennsylvania, PA, June 11–14, 1985, IEEE Cat.# 85CH2127-9, p. 71.
56. J. E. Roederer and G. J. Bastiaans, *Anal. Chem.,* **1983,** *55,* 2333.
57. B. J. Costello, B. A. Martin, and R. M. White, *IEEE Ultrson. Symp. (Proc.),* **1989,** *2,* 977.
58. B. A. Martin, S. W. Wenzel, and R. M. White, *Sens. Actuators,* **1990,** *A21-A23,* 704.
59. A. J. Ricco and S. J. Martin, *Appl. Phys. Lett.,* **1987,** *50,* 1474.
60. Z. Shana and F. Josse, *IEEE Ultrason. Symp. (Proc.),* **1988,** 549.
61. S. J. Martin, A. J. Ricco, G. C. Frye, T. M. Niemczyk, and I. Adhihetty, *IEEE Ultrason. Symp. (Proc.),* **1988,** 607.
62. S. J. Martin, A. J. Ricco, and G. C. Frye, *Sens. Actuators,* **1990,** *20,* 253.
63. J. D. Ferry, *Viscoelastic Properties of Polymers,* 3rd ed., Wiley, New York, 1980.
64. G. C. Frye, S. J. Martin, and A. J. Ricco, *Sensors Mater.,* **1989,** *1,* 335.
65. D. L. Bartley and D. D. Dominguez, *Anal. Chem.,* **1990,** *62,* 1649.
66. S. J. Martin and G. C. Frye, *Appl. Phys. Lett.,* **1990,** *18,* 1867.
67. J. W. Grate, S. W. Wenzel, and R. M. White, *Anal. Chem.,* **1992,** *64,* 413.
68. H. Wohltjen and R. Dessy, *Anal. Chem.,* **1979,** *51,* 1470.
69. D. S. Ballantine and H. Wohltjen *IEEE Ultrason. Symp. (Proc.),* **1988,** 559.
70. D. S. Ballantine and H. Wohltjen in *Chemical Sensors and Microinstrumentation,* R. W. Murray, R. E. Dessy, W. R. Heineman, J. Janata, and W. R. Seitz, Eds., *ACS Symp. Ser.,* **1989,** *403,* 222.
71. J. W. Grate, M. Klusty, R. A. McGill, M. H. Abraham, G. Whiting, and J. Andonian-Haftvan, *Anal. Chem.,* **1992,** *64,* 610.
72. S. J. Martin, G. C. Frye, and S. D. Senturia, *Anal. Chem.,* **1994,** *66,* 2201.
73. D. C. Stone and M. Thompson, *Anal. Chem.,* **1993,** *65,* 352.
74. N. Hill, W. E. Vaughan, A. H. Price, and M. Davies, *Dielectric Properties and Molecular Behaviour,* Van Nostrand Reinhold, London, UK, 1969, Chapter 1.
75. T. M. Niemczyk, S. J. Martin, G. C. Frye, and A. J. Ricco, *J. Appl. Phys.,* **1988,** *64,* 5002.
76. F. Josse, D. T. Haworth, U. R. Kelkar, and Z. A. Shana, *Electron. Lett.,* **1990,** *26,* 834.

77. S. Liew, F. Josse, D. T. Haworth, Z. A. Shana, U. R. Kelkar, and M. Grunze, *IEEE Ultrason. Symp. (Proc.),* **1990,** *1,* 285.
78. J. Andle, J. Vetelino, M. Lade, and D. McAllister, *IEEE Ultrason. Symp. (Proc.),* **1990,** *1,* 291.
79. C. Caliendo, A. D'Amico, P. Verardi, and E. Verona, *IEEE Ultrason. Symp. (Proc.),* **1990,** *1,* 383.
80. F. L. Dickert, G. Bertlein, G. Mages, and W.-E. Bulst, *Adv. Mater.,* **1990,** *2,* 420.

CHAPTER

4

SAW DEVICE OPERATION AND MEASUREMENT

In Chapter 2 we introduced the theory of piezoelectricity, together with the different device formats used for surface-launched acoustic wave sensors. In Chapter 3 we briefly introduced the concept of interdigital transducers, and discussed in detail the mechanisms for perturbation of acoustic velocity and amplitude that can be exploited to transduce chemical and physical phenomena. In this chapter we consider SAW devices from an electrical viewpoint, and examine how we can obtain and measure signals corresponding to physicochemical perturbations of the acoustic wave. To this end, it is useful to consider the equivalent ciruit of the device—that is, a theoretical circuit made up of ideal components that has the same electrical properties and characteristics as the SAW device being modelled. Fundamental to all this is an understanding of interdigital transducer function and properties. It should be emphasized that many different designs and geometries of IDT are possible depending on the precise electrical properties required of the finished device (Fig. 4.1). Since almost all reported SAW chemical sensors employ the same simple IDT geometry, however, discussion of these variations is beyond the scope of this text. Throughout this chapter, therefore, we assume that the IDT consists of alternating polarity (single-tap), equal length fingers (unapodized) having a metallization ratio of 0.5 (equal finger width and gap spacing). For a discussion of the other possible IDT geometries, the interested reader is directed to the ample literature on the subject (1–5).

4.1. INTERDIGITAL TRANSDUCER CAPACITANCE

An IDT can be considered as two coplanar electrodes deposited on the surface a dielectric medium, forming a capacitor. Calculation of the total transducer capacitance (C_T) is possible if the static capacitance for two adjacent fingers (C_s) can be derived. Since the IDT array can be thought of as a number of individual capacitors connected in parallel, C_T is then the sum of all the C_s values. For the single-tap IDT shown in Figure 4.2, simple electrostatic considerations show that the charge distribution on the outermost fingers will be lower than on the inner fingers, resulting in a different value of C_s for the end

Figure 4.1. Different format interdigital transducers.

finger pairs. This is referred to as the "end effect," and is often assumed to be negligible for practical IDTs. For reasons that will become clear later, C_s is defined as a capacitance per unit length. For n fingers (or $N = n/2$ finger pairs), the total transducer capacitance is given by

$$C_T = (2N - 1)C_s W \tag{4.1}$$

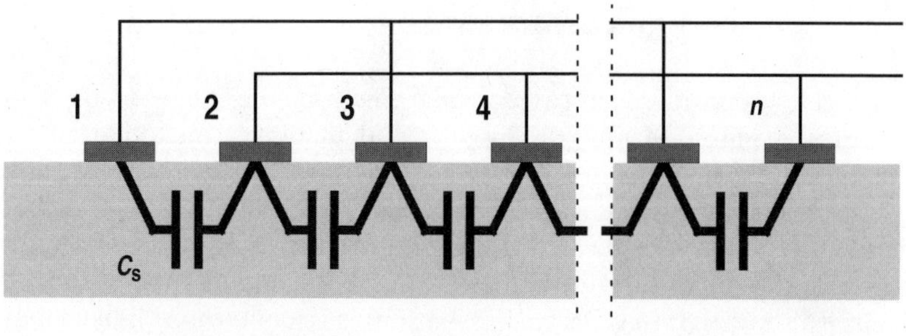

Figure 4.2. The IDT as a distributed static capacitance, C_s.

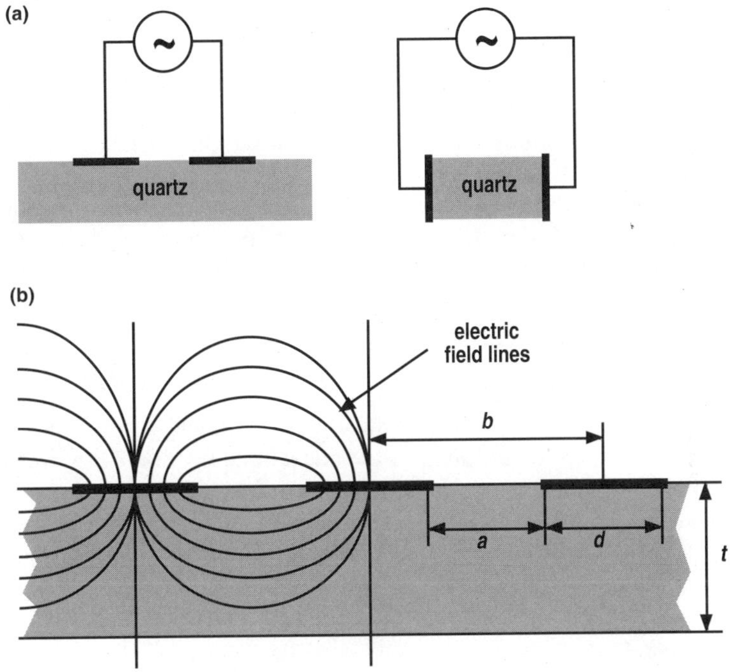

Figure 4.3. (*a*) Relationship between coplanar and parallel plate capacitors. (*b*) Geometrical terms used in the calculation of interdigital capacitance.

where W is the beam width (overlap of adjacent fingers). For a simple parallel-plate capacitor (Fig. 4.3*a*), the capacitance, C, is

$$C = \frac{A\varepsilon_0\varepsilon_r}{l} \tag{4.2}$$

where A is the electrode area, ε_0 is the permittivity of free space, ε_r is the relative permittivity of the dielectric, and l is the electrode separation. The calculation for a coplanar capacitor is obviously more complicated, because now the electric field lines between oppositely charged fingers will be curved through both the substrate and the medium in contact with it (Fig. 4.3*b*). Two approaches exist: (1) contour mapping of the electric field using a finite-element numerical calculation and (2) conformal mapping to derive analytical expressions for the capacitance. Both approaches have been applied since the capacitance of an interdigitated electrode array is of interest not just to the calculation of IDT capacitance (2,6–10), but also to electrochemical, conductance-, and capacitance-based sensor applications as well (11–13). Complicating factors in the case of piezoelectric substrates are the inherent anisotropy of the medium and the inclusion of the dielectric constant in the constitutive relationships. This means that (1) the dielectric constant will vary with substrate deformation and (2) that the substrate dielectric constant must be derived from the dielectric tensor matrix. The static capacitance is therefore calculated using the stress-free permittivity, $\varepsilon_p{}^T$:

$$\varepsilon_p{}^T = (\varepsilon_{yy}\varepsilon_{zz} - \varepsilon_{yz}^2)^{1/2} \tag{4.3}$$

It is also assumed that the IDT height is negligible compared to other dimensions, and C_s is determined as the capacitance per unit length in order to reduce the analysis from three dimensions to two. Further assumptions are often made regarding the dielectric contributions of the substrate and medium in contact with the substrate surface, and on the effect of substrate thickness. One final assumption has to do with the transducer geometry. For two coplanar strips, the capacitance is determined in terms of the whole electrode width, that is, in terms of the near- and far-edge distances (14). However, for two fingers in the middle of the array, only half of each finger is involved in the calculation of C_s (Fig. 4.3*b*). It is therefore assumed that the far-edge distance lies between the finger centers.

The use of conformal mapping to calculate C_s has been clearly described by Kaiser and Castro (6). In this case, the substrate was assumed to have a very much greater dielectric constant than the medium in contact with it, and the electric field was considered largely confined to the substrate. The coordinate system shown in Figure 4.3*b* was first mapped into the complex plane and then, by two Schwarz–Christoffel transformations, into a plane resembling a parallel-plate capacitor. The "area" and "separation" terms of Equation 4.2 are then related to the original geometry through complete elliptic integrals of the first kind:

$$C_s = \varepsilon_{\text{effective}} \frac{\mathbf{K}[\sqrt{1 - k^2}]}{\mathbf{K}[k]} \tag{4.4}$$

where $\varepsilon_{\text{effective}} = \varepsilon_{\text{substrate}}/2$, $\mathbf{K}[m]$ is the complete elliptical integral of the first kind of function m where $0 \leq m < 1$, and k is a function of the gap-to-finger width ratio (Table 4.1). A similar approach was used by Engan (7), although

TABLE 4.1. Geometrical Functions Used in Interdigital Capacitance Calculations[a]

k	Equation	Ref.
$= \dfrac{\tanh\left(\dfrac{\pi a}{4t}\right)}{\tanh\left(\dfrac{\pi d}{2t} + \dfrac{\pi a}{4t}\right)}$	4.4	6
$= \cos\left(\dfrac{\pi}{2}(1 - d/b)\right)$	4.5	2, 7
$= \sin\left(\dfrac{\pi d}{L}\right)$	4.5	8
$= \dfrac{a}{b}$	4.4	10, 15

[a] See Figure 4.3*b* for definition of geometrical terms.

in this case the geometrical parameter was expressed in terms of the metallization ratio, $r = d/b$. In this case, the numerator and denominator in the second term of Equation 4.4 are reversed, that is,

$$C_s = \varepsilon_{\text{effective}} \frac{\mathbf{K}[k]}{\mathbf{K}[\sqrt{1 - k^2}]} \tag{4.5}$$

but the results are essentially the same. Similar results have been obtained by Auld (2,8), Scapple (10), and Endres and Drost (15). Table 4.1 summarizes the terms used for k by each of these authors. Typical results for C_s as a function of r are shown in Figure 4.4. In a previous study (16) it was reported that very different results were obtained on plotting the calculated C_s values against r. It has since been found that these results were influenced by the method used for solving the elliptic integral functions. Originally, a series expansion of $\mathbf{K}[m]$ was employed (17), truncated after the term in m^3:

$$\mathbf{K}[m] = \frac{\pi}{2}\left\{1 + \left(\frac{1}{2}\right)^2 m + \left(\frac{1 \cdot 3}{2 \cdot 4}\right)^2 m^2 + \left(\frac{1 \cdot 3 \cdot 5}{2 \cdot 4 \cdot 6}\right)^2 m^3 + \cdots\right\} \tag{4.6}$$

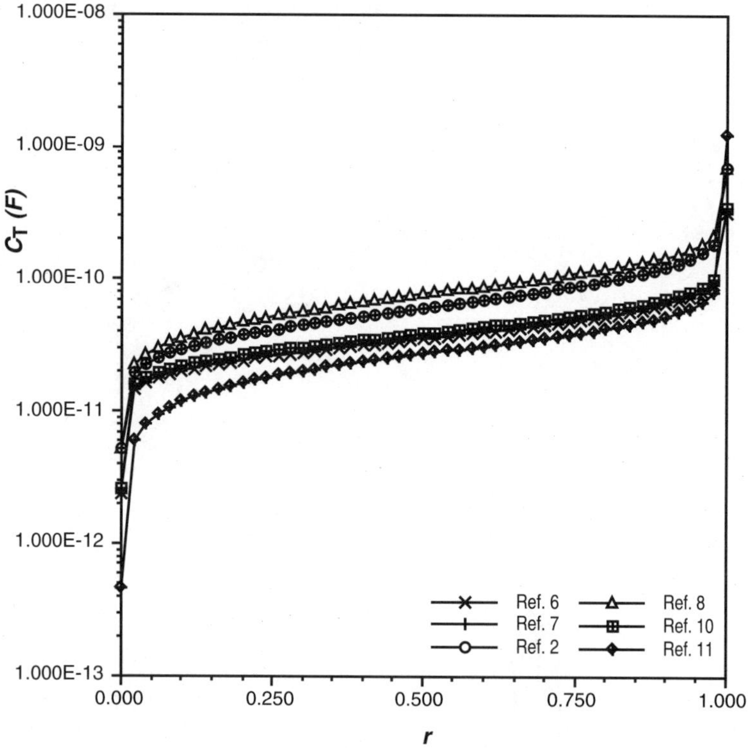

Figure 4.4. Total capacitance of a 50 finger-pair IDT on 2-mm thick ST-quartz with $d = a = 15\ \mu$m and $W = 1.2$ cm as a function of metallization ratio according to the equations in Table 4.1.

In the calculations for Figure 4.4, however, an implementation of Carlson's method was used (18) to provide more accurate results. It is believed that the results described previously were distorted by failure to include sufficient terms in the evaluation of Equation 4.6. In Carlson's method, an iterative procedure is used in which the error reduces by 4^6 for each iteration; 16-digit precision was obtained in as few as 5 or 6 iterations.

In comparing the treatments described in Table 4.1 and Figure 4.4, a major difficulty is the different systems used to denote the transducer geometry, and hence the exact form of the function k. The other difficulty with Equations 4.4 and 4.5 is solving the elliptic integrals. For this reason, Binotto and Piancentini (11) used an approximate formula, Lin et al. (19) used a

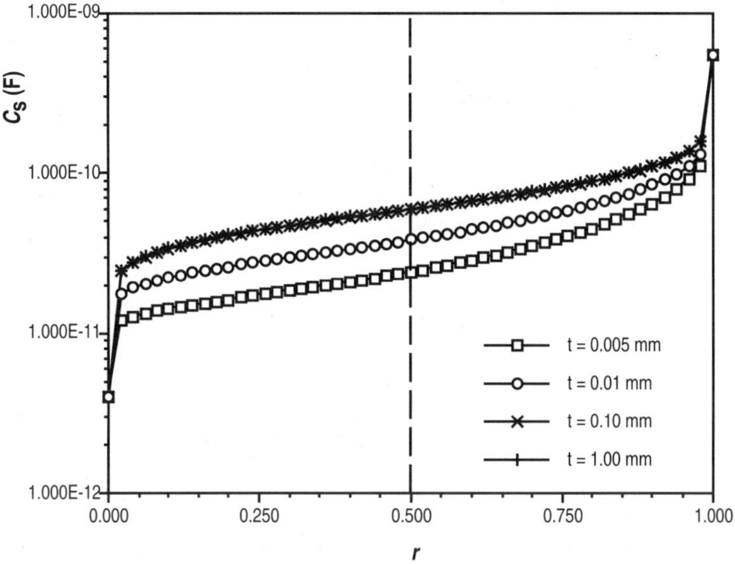

Figure 4.5. Static capacitance of a 50 finger-pair IDT with $b = 30\ \mu$m and $W = 1.2$ cm as a function of r for different substrate thicknesses according to Equation 4.4 (Kaiser and Castro, ref. 6). Substrate dielectric taken as 9.4×10^{-11} F m^{-1}.

finite element calculation, and Duclos and Leduc (20) used a Fourier series calculation. In all cases, similar results were obtained: C_s increases asymptotically as r approaches unity and tends to zero as r tends to zero. Actual discrepancies in C_s at a specific r may be attributed to differences in the treatment of $\varepsilon_{\text{effective}}$ and in the geometrical representation of the transformed coordinates in k. In terms of substrate thickness, C_s increases to a maximum at all r values, representing the maximum penetration of the electric field into the substrate (Fig. 4.5). For all practical Rayleigh wave devices, the substrate thickness is such that it can be considered infinite in terms of calculating C_s. The presence of a ground plane on the reverse side and a dielectric coating on the substrate surface will also both significantly affect C_s (9,20).

4.2. IDT FUNCTION ON PIEZOELECTRIC SUBSTRATES

So far we have treated the IDT solely as a capacitance, neglecting the piezoelectric nature of the substrate. In reality, of course, a voltage applied across the terminals of an IDT results in a mechanical displacement of the substrate. Similarly, mechanical displacement of a short-circuited IDT induces a current through the electric potential wave accompanying the SAW, as described in Section 3.5. To better appreciate the significance of this, we present here a phenomalogical description of IDT function. Consider first the application of a point charge to the surface of a piezoelectric substrate. The resulting dielectric polarization of the substrate will result in displacement of

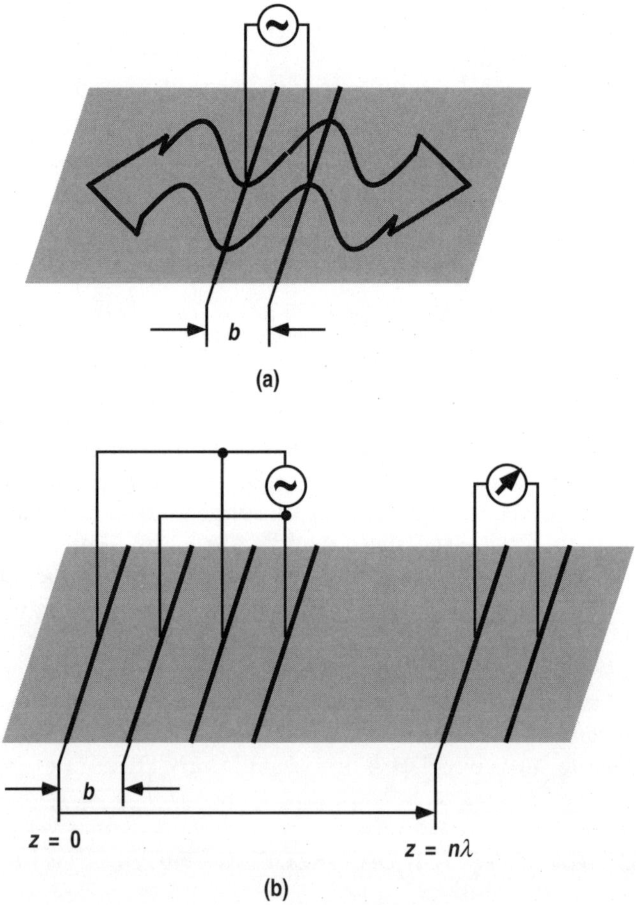

Figure 4.6. (*a*) Acoustic waves propagating from two oppositely charged strip conductors. (*b*) Detection of acoustic waves by a short-circuited IDT at an arbitrary distance z from the generating transducer array.

the atomic centers, giving rise to a surface mechanical deformation. Removal of the charge allows the surface to relax to its former state, resulting in a ripple spreading out from the center with rapidly decreasing amplitude. A line charge will result in two wavefronts propagating in opposite directions; similarly, an oscillating line charge will give rise to acoustic waves travelling in each direction. The deformation produced by adjacent fingers in an idealized IDT is slightly different because the charge (voltage) is applied across a separation b (Fig. 4.6*a*). The result is the same, although now the wavelength is determined by the separation, as noted in Section 3.1:

$$\lambda = 2b \tag{4.7}$$

Consider the current generated in a short-circuited IDT by the converse piezoelectric effect at an arbitrary distance z from the generating IDT. It is apparent that this will be exactly in phase with the generating voltage when $z = n\lambda$, where $n = 1, 2, 3 \ldots$ for a specific frequency $f_0 = v_0/2b = v_0/\lambda$ (Eq. 3.2). This is of particular importance when we consider the general case of the N finger pair IDT with periodicity b commonly used in SAW chemical sensors (Fig. 4.6*b*). The generating voltage is now simultaneously applied to all finger pairs with $z = \lambda$. At f_0, the acoustic wave generated by the first finger pair is thus exactly in phase with that generated by the second pair and so on.

This raises the question of what happens when $f \neq f_0$, since the finger spacing is fixed. First, it should be noted that given the high frequencies of typical SAW devices, small variations in f around f_0 do not give large changes in λ for a fixed v_0. As an example, for a 52.0-MHz device, a change of 1 kHz is needed to achieve an 0.001 μm change in wavelength which, given finger width and gap spacings of 15 μm, is not significant. In other words, the acoustic wave will still be propagated by the structure, although the waves generated by the outermost finger pairs will obviously be out of phase, causing some destructive interference (and hence attenuation) which will increase for increasing N. As f continues to move away from f_0, the wavelength corresponding to the Rayleigh wave velocity (v_0) becomes significantly different from the physical dimensions of the IDT and so propagation of this particular mode will cease. For acoustic wave generation to continue, therefore, we now require an acoustic mode having a different propagation velocity and/or a different wavelength launched by the *same finger spacing*. To accomplish the latter, propagation of the acoustic wave must occur at a different orientation to the surface (Fig. 4.7). In either event, it can be seen that an IDT will

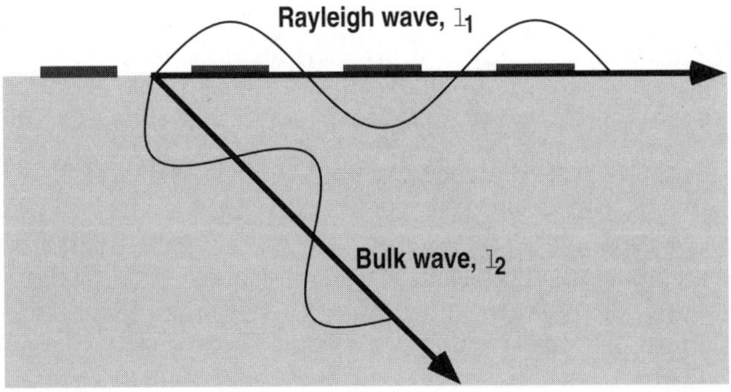

Figure 4.7. Use of an IDT to launch either Rayleigh or bulk waves, having different wavelengths.

generate (or receive) acoustic waves only over specific frequency ranges, and that these ranges decrease in width with increasing N. The same considerations apply to an IDT acting as an acoustic receiver.

Further discussion of IDT properties at this point is often made with reference to the frequency $[H(f)]$ or impulse $[h(t)]$ response, which consti-

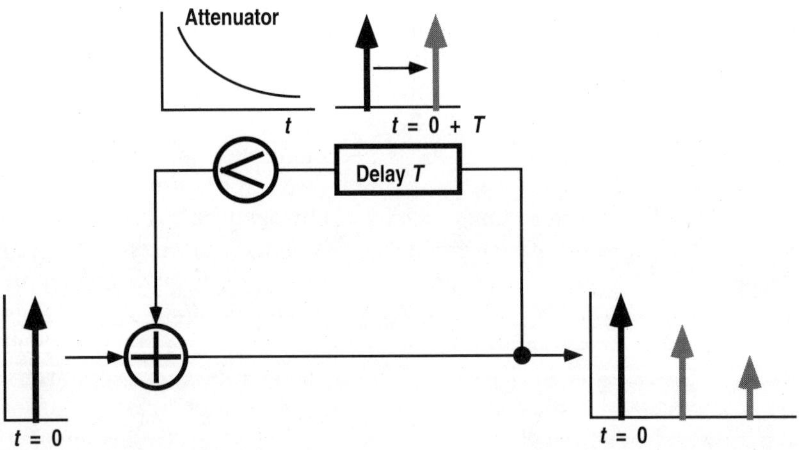

Figure 4.8. Schematic diagram of a transversal filter incorporating a delay and an attenuator, with resulting impulse response function.

tute a Fourier pair (4):

$$H(f) = \int h(t) \exp(-j\omega t)\, \mathrm{d}t \tag{4.8}$$

$$h(t) = \int H(f) \exp(+j\omega t)\, \mathrm{d}t \tag{4.9}$$

where $\omega = 2\pi f$. The frequency response of an IDT can thus be discussed in terms of the magnitude and phase of $H(f)$. From the preceding phenomenological description of IDT function and the consideration of piezoelectric theory outlined in Chapter 2, it can be appreciated that the impulse response of an IDT can be modeled by a circuit such as that in Figure 4.8, in which the initial pulse is delayed and attenuated before being fed back into the circuit. This constitutes a transversal filter, for which it can be shown from sampling theorem that the frequency response for large N is (4)

$$H(f) = \int_0^{NT} \exp(-\,j\omega t)\, \mathrm{d}t = \frac{N}{f_0} \exp(-\mathrm{j}N\pi f/f_0) \frac{\sin(N\pi f/f_0)}{N\pi f/f_0} \tag{4.10}$$

where $T = 1/f_0$. It can then be shown that the phase of $H(f)$ varies linearly with frequency while the magnitude is as shown in Figure 4.9. Note that, because the IDT fingers sample the acoustic wave at discrete intervals rather than continuously, the actual frequency response will consist of multiple passbands centered around $\pm nf_0$ where $n = 1, 2, 3 \ldots$. This has the effect that the side lobes will be much narrower than the main passband at nf_0.

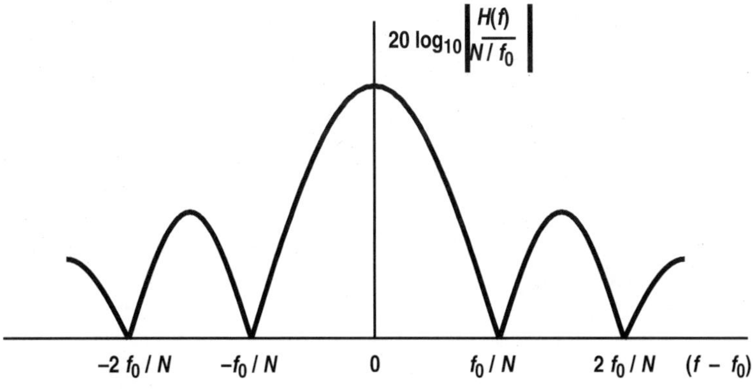

Figure 4.9. Plot of the magnitude of $H(f)$ for an IDT.

4.3. EQUIVALENT CIRCUIT OF AN IDT

We now consider the nature of the IDT in terms of its electrical properties, as represented by an equivalent circuit. We have so far discussed the generation of acoustic waves by the IDT fingers, and the mechanical interaction of these waves with one another as they propagate in both directions from each finger pair. These waves also interact with the IDT fingers, however, giving rise to surface wave reflections within the IDT. There are two causes of this phenomenon (21): (1) the metal electrodes contribute additional surface mass and short circuit the electric potential wave accompanying the SAW (Section 3.5), resulting in different acoustic impedances for the electrode and gap regions; and (2) this same potential wave delivers a voltage to the electrical load, resulting in a "regeneration" of the surface waves. Other important factors have to do with the piezoelectric properties of the substrate. The induced strain will, depending on the applied frequency relative to the mechanical resonance frequency, either lead or lag the applied voltage (energy is either stored or lost). At mechanical resonance, the strain is exactly in phase with the applied voltage, and only frictional losses are involved (22). We therefore have to consider two aspects of the IDT: its electrical (capacitance) and acoustic properties, and the interconversion of electrical and acoustic energy.

The Mason model (Fig. 4.10*a*) is often used to describe the IDT, since this incorporates both aspects of IDT behavior and allows calculation of the electrical insertion loss. This parameter is of importance to both resonator and delay-line oscillators, as we shall see later. The Mason model (23) was originally developed for use with bulk-wave trandsucers such as the thickness shear mode disc resonator. This models the device as a transmission line (i.e., a circuit made up of repeating units having constant impedance per unit length) made up of separate components representing the acoustic and electrical properties. These repeat units are cascaded acoustically and connected in parallel electrically to describe an *N*-finger pair IDT.

There are two variants of the model as applied to SAW devices: the crossed and in-line field models (24). These terms refer to whether the electric field is taken as perpendicular or parallel to the direction of SAW propagation, and provide good approximations to IDT behavior. The choice of model depends on whether the particular substrate used shows better acoustoelectric coupling to the perpendicular or parallel electric field components associated with the IDT geometry. Further refinements of the model are also available (1). While often cited in the SAW chemical sensor literature, the Mason model is, however, not strictly necessary to an understanding of the electrical aspects of IDTs for our purposes. It should also be noted that it is possible to

(a)

(b)

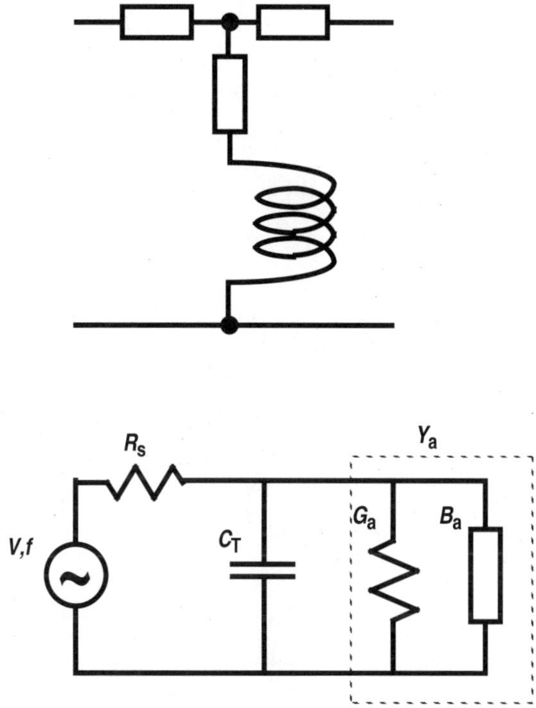

Figure 4.10. (*a*) Mason model for an IDT. (*b*) Simplified equivalent circuit for an IDT showing the source resistance (R_s), transducer capacitance (C_T), and the acoustic admittance (Y_a).

derive IDT insertion loss and transfer characteristics (important in the design of SAW filters) from transfer functions defined without the aid of an equivalent circuit (4). For this reason, we use here a simplification of the cross-field Mason model consisting of the IDT capacitance and an admittance arising from acoustic interactions with the IDT fingers as described above [a series arrangement better corresponds to the in-line field model (24)].

The equivalent circuit for an IDT functioning as an acoustic wave generator is shown in Figure 4.10*b*. This consists of the voltage source (V), source impedance (R_s), IDT capacitance (C_T), and the complex admittance (Y_a) composed of the radiation conductance (G_a) and susceptance (B_a):

$$Y_a = G_a + jB_a \tag{4.11}$$

where the subscript "a" denotes an acoustic origin. Here G_a and B_a are Hilbert transforms of one another, allowing the susceptance to be calculated from the conductance and vice versa. Since parallel admittances are additive, the net admittance of the IDT is, therefore,

$$Y_{\mathrm{IDT}} = Y_{\mathrm{T}} + G_{\mathrm{a}} + B_{\mathrm{a}} \tag{4.12}$$

where $Y_{\mathrm{T}} = j\omega C$. Although a full derivation is beyond the scope of this text, it can be shown that

$$G_{\mathrm{a}} = G_{\mathrm{a}}(f_0)\frac{\sin^2 x}{x^2} \tag{4.13}$$

$$B_{\mathrm{a}} = G_{\mathrm{a}}(f_0)\frac{\sin(2x) - 2x}{2x^2} \tag{4.14}$$

where $x = N\pi(f - f_0)/f_0$ and $G_{\mathrm{a}}(f_0)$ is a constant that measures the effectiveness of acoustoelectric coupling at the fundamental resonance frequency of the IDT. Note that Equations 4.13 and 4.14 follow from Equation 4.10,

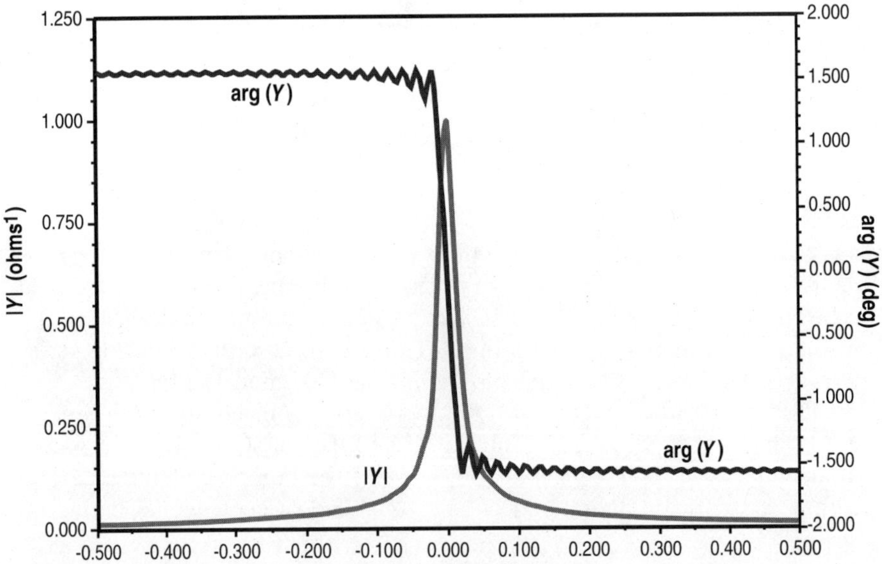

Figure 4.11. Magnitude and phase of the admittance for an IDT.

with the modification that the terms are measured around the center frequency f_0. The radiation conductance accounts for the electrical power from the voltage source carried by the acoustic wave away from the IDT. The radiation susceptance accounts for the currents induced by acoustic waves passing beneath the fingers of the generating IDT. Since the finger pairs are exact multiples of λ apart, at f_0 this generated current will be in phase with the applied voltage and B_a will be zero. At other frequencies, however, the current will either lead or lag the voltage and so B_a will be either capacitive (positive) or inductive (negative), respectively. Figure 4.11 shows the relative magnitude and phase of the IDT acoustic admittance calculated from Equations 4.13 and 4.14 for a 50-finger pair, 60-μm wavelength IDT on an ST-X quartz substrate. The impedance characteristics of the IDT may be inferred from Figure 4.11 because impedance and admittance are the reciprocal of one another and, for such a relationship, the magnitude and phase are such that $|Z| = |Y|^{-1}$ and $\arg(Z) = -\arg(Y)$.

4.4. DEVICE DESIGN AND FABRICATION

SAW devices are manufactured using lithographic techniques on optically flat substrates. The choice of substrate is largely determined by the application and desired device characteristics. From an electrical viewpoint, the key factors are the piezoelectric coupling coefficient, K^2, and the temperature coefficient. These vary between piezoelectric materials and different orientations of the same material. Lithium niobate ($LiNbO_3$) is widely used because it has the highest K^2 value among piezoelectric substrates capable of supporting high-frequency SAWs (Table 4.2). Quartz has a much lower K^2 value ($\approx 1/40$ th that of $LiNbO_3$), but this can be overcome to a large extent by

TABLE 4.2. Material Constants for Different Piezoelectric Substrates

Substrate	K^2	C_s (pF cm^{-1})	v_0 (m s^{-1})	Ref.
ST-X quartz	0.0011	0.5	3158	4, 25
Y-X quartz	—	—	3159	2, 26
128°Y-Z $LiNbO_3$	0.048	4.6	3488	4, 25
77.5°Y-90°X $LiTaO_3$	0.016	4.4	3379	4, 25
100–011 GaAs	0.0007	1.2	2864	4, 25
Y-60°X CdS	—	—	1702	2, 26
ZnO	—	—	2639	2, 26

increasing the number of finger pairs in the IDTs. One highly attractive feature of ST-cut, X-propagating quartz is its zero-temperature coefficient around room temperature. Quartz also has a greater mass sensitivity than $LiNbO_3$, and allows a greater variety of cleaning and coating strategies to be implemented. The latter include covalent attatchment of mono- and multi-layer structures for improved sensor lifetime and stability (Chapter 5).

Other substrate materials employed for electronic device fabrication include lithium tantalate and gallium arsenide. Layered $ZnO-SiO_2-Si$ structures have also been described for both Rayleigh wave (27–30) and flexural plate-wave (31,32) chemical sensors. These offer the potential for monolithic integration of the SAW sensor with electronic signal-processing circuitry, since they are manufactured on silicon substrates using the same techniques. As described briefly in Chapter 2, the silica layer ($\approx$1 μm) can be grown directly on the silicon substrate by thermal oxidation in an oxygen atmosphere. A thin metal layer (such as aluminum) may then be deposited, followed by r.f. sputtering of the ZnO layer from a zinc metal target in an oxygen plasma. The ZnO layer may be from 1 to 20 μm thick for Rayleigh wave and $<$1 μm for FPW devices, which may also be fabricated using silicon nitride ($\approx$2 μm) instead of silica. If a metal layer is included in the structure, the IDTs are formed on the upper surface. Alternatively, the IDTs may be formed on the silica surface with the ZnO layer on top. Finally, certain polymeric piezoelectric materials are available including poly(vinylidene fluoride) (33).

Optically flat substrates are typically obtained using a combination of mechanical and electron beam polishing. The substrate would then be coated with a suitable resist. A mask is prepared from a scale drawing of the required device, including the IDTs and any required gratings, reflectors, contact points, and grounding strips. Optical lithography is used to transfer the image to the resist, which is then developed to form a stencil. Metallization is then achieved by thermal evaporation or sputtering. Aluminum is preferred from an electrical viewpoint owing to its elasticity; gold is preferred from a chemical viewpoint owing to its resistance to chemical attack. Gold does not adhere well to the substrate surface, however, so a thin adhesion layer of chromium metal is employed. Metallization typically involves the deposition of $\approx$1000 Å thick films, and is followed by removal of the stencil and unwanted metal. In some cases, the finished device is coated with a thin silicon dioxide layer to provide a scratch-resistant coating over the IDTs.

Clearly, the production of SAW devices requires access to microelectronic fabrication facilities which may not be available to the chemical sensor researcher. There are, however, a number of commercial sources for both devices and complete sensor systems, although it should be noted that these may not include some of the more recent structures such as the FPW device.

TABLE 4.3. Companies Supplying SAW Devices and Systems

Company	Product(s)
Crystal Technologies, Inc. (Palo Alto, CA)	Substrates, devices
Femtometrics, Inc. (Irvine, CA)	Devices
Hewlett-Packard (Palo Alto, CA)	Devices
Microsensor Systems, Inc. (Bowling Green, KY)	Devices, systems, vapor generators
Plessey Semiconductors (Towcester, UK)	Devices
SAWTEK (Orlando, FL)	Devices
Xensor Integration bv (Delft, Netherlands)	Devices, systems
Valpey-Fisher (Hopkinton, MA)	Substrates

A selection of those companies that have been cited in the chemical sensor literature is included in Table 4.3.

There are two general categories of SAW device: resonators and delay lines (Fig. 4.12). In the resonator configuration, acoustic waves are launched within a resonant cavity such that the signal is reflected back to the generating IDT(s). Resonators may be either one- or two-port devices, that is, having one or two IDTs deposited onto a piezoelectric substrate between a pair of grating reflectors (Fig. 4.12*a*). Delay lines are always two-port devices having two IDTs deposited on a piezoelectric substrate with between-centre separation equal to a whole number of wavelengths (Fig. 4.12*b*).

The final dimensions of SAW devices are a function of the desired operating frequency, and can be determined by employing various scaling laws. We have already seen that the finger width and spacing required for a given f_0 value can be calculated from the propagation velocity of the acoustic mode employed. In addition, the bandwidth (range of frequencies centred around f_0 which will be passed by the transducer) is determined by the number of finger pairs. An expression for the bandwidth can be obtained from the transducer response characteristics (Section 4.2). For attenuations $\leq$ 3 dB* the bandwidth is

$$f_b = \frac{0.88 f_0}{N} \tag{4.14}$$

*Attenuation is commonly expressed in decibels, which are a log ratio of either voltage or power relative to a reference level. For example, 0 dBm is defined as 1 mW power into a 600-Ω load while 0 dBv is defined as 0.775 V into a 600-Ω load. For voltages, attenuation is calculated as 20 $\log_{10}(V_0/V_i)$ and for power as 40 $\log_{10}(P_0/P_i)$.

(a)

(b)

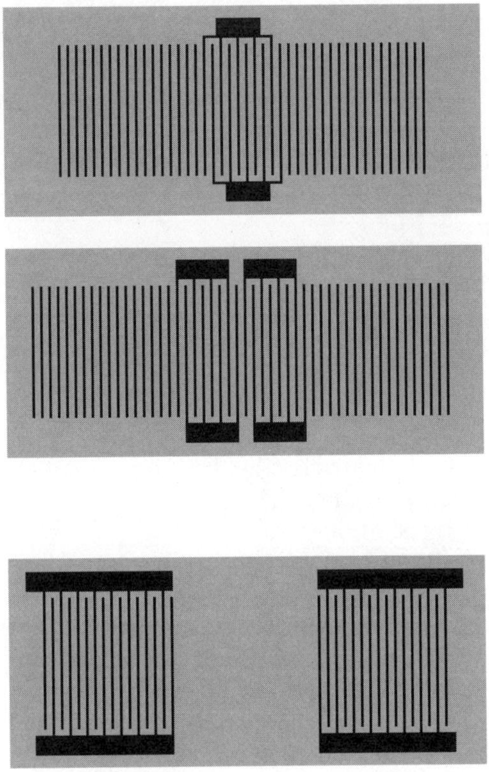

Figure 4.12. (*a*) One- and two-port SAW resonators. (*b*) Two-port SAW delay line.

Thus the minimum insertion loss for a SAW delay line will be 6 dB. Larger bandwidths can be accommodated without increased insertion loss on $LiNbO_3$ substrates compared to quartz (34).

Other losses will arise from spurious signals such as "triple transit echo" and interaction of the SAW with bulk waves simultaneously generated by the IDT and reflected from the back face of the device (34). A triple-transit echo arises from the bidirectional nature of the IDTs, and occurs when the waves propagating away from the center of the delay line reflect off the physical edge of the substrate. This problem obviously does not arise for resonators, but is an important consideration for delay lines, and may be addressed in several ways: (1) a strip of an acoustically absorbing material can be deposited on each end of the substrate, (2) the substrate may be cut with the ends angled to reflect the unwanted signal out of the path of the delay line, or (3) by using unidirectional transducers. Bulk-wave effects may be reduced by leaving the back surface of the substrate rough, by coating it with an acoustic absorber, or by imposing a grating structure on it.

Another important consideration is the propagation length (L) between IDT centers. Attenuation (or device impedance) increases with increasing L, so that shorter delay lines would be preferred on the basis of insertion loss only. Noise levels, however, are reduced by increasing L. This observation follows from the condition for an oscillator circuit (34):

$$\frac{\omega L}{v} + \Phi_E = 2n\pi, \qquad n = 0, 1, 2, 3, \ldots \tag{4.15}$$

where Φ_E is the phase shift associated with the electronic circuit and $\omega L/v$ is the phase shift associated with the acoustic wave. Since the main source of random noise is small phase shifts ($\delta\Phi_E$) in the electronic circuit, the resulting frequency deviation (δf) can be expressed as

$$\frac{\delta f}{f} = -\frac{\delta\Phi_E}{2\pi N_\lambda} \tag{4.16}$$

where $N_\lambda = L/\lambda$. In fact, calculations suggest that the optimum length for a 100-MHz delay line on ST-cut quartz is 1 m, which is highly impractical given that the largest available substrates are 25 cm long. A reasonable tradeoff is to pick a value of L such that the resulting frequency noise is $\leq$1 Hz. Note that SAW resonators have an implicit advantage here, since the effective value of L is a function of the reflectivity. Impedance matching with the electronic circuit is also important (we discuss this in more detail in the following section), which leads to the choice of beam width (W) as an integer multiple of λ. This leads to the observation by Wohltjen (26) that device length and width scale inversely with f_0, while device area scales inversely with f_0^2. The implications of this for analytical sensitivity are dealt with in Chapter 5. Practical limitations on f_0 for a given substrate are therefore established by (1) the largest delay length that can be accommodated (lower-frequency limit) and (2) the spatial resolution of the techniques used in forming the IDTs (upper-frequency limit).

4.5. SAW SENSOR OPERATION

Having examined the functioning of the IDT in some detail, we are now in a position to discuss the operation and configuration of the various SAW devices employed as chemical sensors. In general terms, these exploit the electrical properties of the IDT as discussed above, together with the inherent delay produced by the much lower propagation velocity (longer wavelength) of an acoustic rather than a purely electrical signal of the same frequency. For both resonators and delay lines, perturbation of v_0 through the mechanisms outlined in Chapter 3 will result in a phase shift between the applied and received (or reflected) signals such that the frequency for minimum impedance is shifted away from f_0. Mechanical damping of the acoustic wave will also result in a decrease in the amplitude of the received (or reflected) signal. In other words, perturbation of the acoustic wave results in both attenuation and phase-shifting of the electrical signal. Changes in v can therefore be measured through changes in resonance frequency, phase shift, electrical impedance, and attenuation (insertion loss) of the device.

These parameters can be measured in two basic ways. Either the device can be incorporated as the frequency-controlling element of an oscillator circuit (the "oscillator method"), or an external frequency signal may be applied and the phase and amplitude compared with the transmitted or reflected signal. The former is more common because the equipment requirements to obtain a measurement are much simpler and less expensive. The latter are of particular use in determining device characteristics and elucidating acoustic perturbation response mechanisms. A variation on the direct phase–amplitude measurement is the use of pulse echo techniques, in which a voltage pulse is applied, as opposed to continuous excitation. In this respect, acoustic waves may also be generated by laser pulse excitation. All of these techniques have been employed by different authors at various times within the SAW chemical sensor literature and are discussed in detail in the following sections.

4.5.1. Oscillator-Based Measurements

The generalized circuits for incorporating one- and two-port SAW devices into an oscillator circuit are shown in Figure 4.13. The SAW device, having a net impedance Z_{SAW}, is connected in the feedback path of an amplifier with gain A. The resistance R includes the input resistance of the amplifier and, together with Z_{SAW}, forms a voltage divider that determines the feedback fraction, β (Fig. 4.13*a*). The instantaneous feedback (v_f) and output (v_0) voltages are therefore related by

$$v_f = \beta v_0 = \beta A v_i \tag{4.17}$$

A small voltage appearing at the input is amplified and fed back through the SAW device from the output. The SAW device will delay this signal by a certain amount before it appears at the amplifier input, subsequently reappearing at the SAW input terminal. The result will be an alternating signal that will be self-sustaining provided that, at the frequency determined by the

(a)

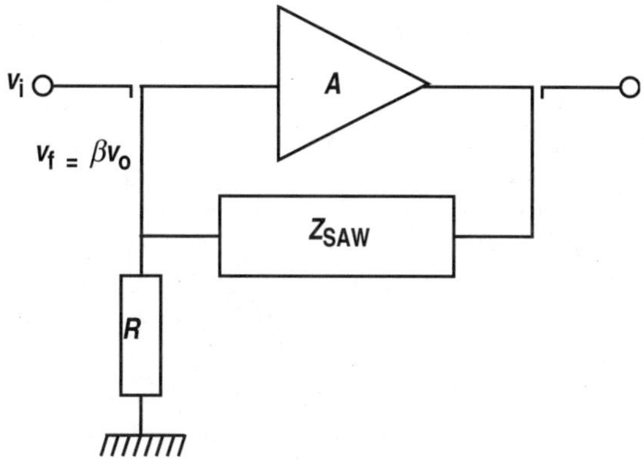

(b)

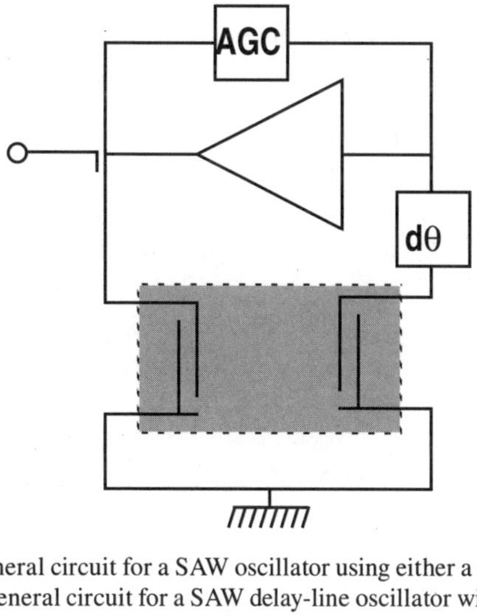

Figure 4.13. (*a*) General circuit for a SAW oscillator using either a resonator or delay line. See text for details. (*b*) General circuit for a SAW delay-line oscillator with AGC and phase control.

SAW device, v_f is equal to v_i. By inspection, we can see that Equation 4.17 is satisfied when

$$\beta A = 1 \tag{4.18}$$

This is the loop-gain equation. Since the feedback fraction β depends on R and Z_{SAW}, and Z_{SAW} is complex, we can see that for Equation 4.18 to be satisfied, the magnitude of the loop gain must be unity and the phase of the loop gain must be an integer multiple of 2π (Eq. 4.15). If $|\beta A| < 1$, the signal will decay away; if arg $(\beta A) \neq 0$, v_i and v_0 will be out of phase, and destructive interference will occur, resulting in a loss of oscillation.

In real circuits, $|\beta A|$ may be greater than unity. Under these circumstances, the signal will continue to increase in magnitude until the practical limitations of the amplifier and power supply are met (the amplifier is said to "saturate"), and a maximum output level is attained. It is possible, however, to achieve a marginal oscillator in which the amplifier gain (or the feedback fraction) is adjusted so that the loop gain is always unity. This is known as automatic gain control (AGC), and has the advantage that the AGC level gives a direct measure of changes in SAW attenuation due to mass and/or viscoelastic loading. Another feature of real circuits is that the amplifier and other electrical components will all have finite phase properties, some of which may not be negligible. To maintain the loop-phase condition under all SAW perturbation conditions, therefore, a variable phase delay can be incorporated into the circuit (Fig. 4.13*b*). This also has the advantage of allowing variations in the phase of the SAW device to be monitored in addition to the oscillation frequency.

While frequency is the primary measurement in many practical SAW chemical sensor applications, phase and attenuation are also useful parameters, especially in the context of viscoelastic loading, as we saw in Section 3.7. Practical oscillator circuits for SAW delay lines and two-port resonators have been published by Parker (35), Wohltjen (26), Rezgui and Alder (36), and Chang et al. (37), although very few circuit details are provided in the latter reference. Phase and amplitude measurements can also be made using using the AGC and phase-delay systems described above. Suitable circuits have been described briefly by Hietala et al. (38). Oscillator circuits made from discrete components (rather than a dedicated circuit board) have also been described (39,40). In the latter case, cabling and impedance issues become especially important. The main features required of a suitable circuit are a broadband, noninverting r.f. amplifier, a means of monitoring the oscillation frequency, and some form of impedance matching network. A stable power supply is also essential. A low-pass filter whose cutoff frequency is above the SAW resonance frequency can also be included in the feedback loop to eliminate spurious oscillation at higher frequencies and mode hopping. The latter occurs whenever the attenuation of the fundamental mode (i.e., Rayleigh wave propagation) is such that a lower circuit impedance is obtained for a different acoustic propagation mode generated at a different frequency, as discussed above (Section 4.2). The result of this can be sudden large jumps in baseline frequency and abrupt changes in the direction of frequency change on vapor sorption (16).

At the high frequencies employed with SAW devices (10–500 MHz), electrical signals propagate more as electromagnetic waves "guided" by the conductor. Any length of conductor can act as a transmitter or receiver of r.f. signals. Furthermore, if such a signal propagating along a cable sees a change in electrical impedance, either along the cable length or at a connection, the signal will become attenuated through phase and signal-reflection effects. It is therefore essential to match the input and output impedances of all the components within the oscillator circuit. In the case of discrete components, the cables must also be of the same constant impedance—normally, coaxial cables having a 50-Ω impedance are used. A related effect is that careful consideration must also be given to the cable length. Radio-frequency signals will propagate along a coaxial cable at a certain fraction of the speed of light—typically around 66%. For a 60-MHz signal this corresponds to a wavelength of 3.3 m. If the total length of cable in the oscillator loop is significant, the cabling will itself contribute a phase shift to the signal, resulting in interference. When the cable length approaches an odd multiple of half wavelengths, the signals in and out of the amplifier will be 180° out of phase and will cancel, preventing oscillation. If for physical reasons the electrical connections cannot be kept short (say less than 2% of the wavelength or 7 cm in our example), the total length of cable should be made close to the wavelength to prevent this situation from arising (41,42). In dedicated oscillator circuits, this problem is readily overcome by using a double-sided circuit board, with the SAW device on one side and the oscillator circuitry on the other.

Impedance matching of the electrical circuit to the SAW transducers is important whether a circuit board or discrete components are used. The situation is complicated by a number of factors, including the effect of chemical and physical perturbation of the SAW velocity as described in Chapter 3. The simplest approach described in the literature is to use balancing transformers (43) or r.f. chokes (inductors) (26,42,44) to cancel the contribution of the IDT static capacitance. A 50-Ω resistor is also sometimes connected in parallel across each IDT to provide the required load impedance (26). More rigorous approaches use proprietary, custom-designed impedance-matching networks (40).

Other impedance-matching strategies also incorporate tuning components to allow "pulling" or "trimming" of the SAW oscillator frequency (4,35). This is more important in bandpass filter applications, but also of relevance to SAW chemical sensor implementations. Most SAW sensor applications employ a difference frequency system, where two SAW oscillators are used to provide some degree of temperature and pressure compensation (Section 4.6). The two SAW devices may be either physically separate or manufactured on the same substrate. Unfortunately, if the frequencies of the two oscillator circuits are close (typically within 1 kHz for 52-MHz devices), the two can "lock" together and oscillate at identical frequencies under all conditions. This is especially a problem when the SAW devices are physically located on the same substrate owing to acoustic cross-talk between them. However, similar problems can arise due to electrical cross-talk between the circuits as a result of the high frequency of operation. Acoustic cross-talk can be reduced by using acoustic absorbing polymer beads on the device surface and metal grounding strips (25,45). Another approach is to tune one oscillator circuit away from the other by incorporating a variable capacitor in parallel with one IDT. Under certain circumstances, however, this can also result in anomalous frequency shifts and excessive baseline noise (16).

4.5.2. Phase/Amplitude-Based Measurements

Figure 4.14 shows various connection schemes for measuring the impedance properties of a two-port SAW device. These can be performed using a network analyzer or frequency source and vector voltmeter combination. Examples of phase and amplitude (or attenuation) measurements include the original work of Wohltjen and Dessy (46), Falconer et al. (47), and Martin

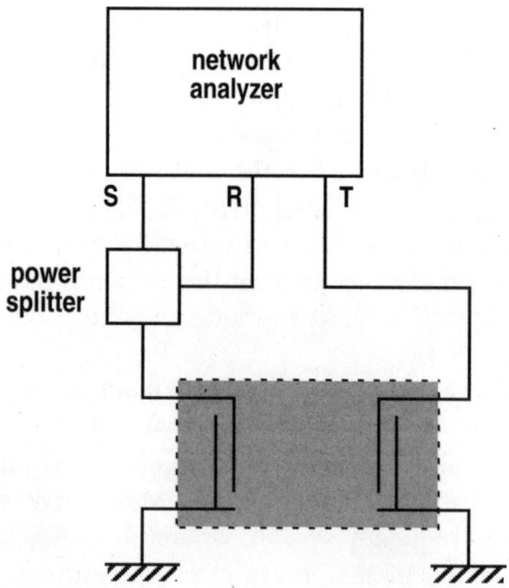

Figure 4.14. Use of a spectrum–network analyzer for the measurement of SAW delay-line properties.

et al. (48). In all these examples, an external signal source is applied to one IDT of the SAW device and a load impedance to the other. When this is done, it is first necessary to determine the optimum frequency to which to tune the source. This can be determined using a network (or impedance) analyzer to determine the frequency of minimum attenuation. It would also be possible to connect a vector voltmeter across the input and output of a SAW in an oscillator circuit, but using the signal generator gives much better control over both the frequency and amplitude of the input voltage. The vector voltmeter allows both the phase shift and attenuation change (insertion loss) due to acoustic wave perturbation to be monitored.

Network analysis has been used by various authors, including D'Amico et al. (49), Chang et al. (37), and Inoue et al. (50). In this technique, a signal of fixed power is applied at different frequencies, and the ratio of the input to transmitted power measured. This allows both the magnitude and phase of the electrical impedance to be measured. It is also possible to fit different equivalent circuits to this data using complex multiple linear regression analysis (51). The advantage of network analysis is that a more complete characterization of the device around the resonance frequency is possible, even under conditions of fluid or viscous loading that would prohibit use of oscillator-based measurements. The ability to fit different equivalent circuit models to the impedance data is also beneficial in the development of theoretical models for SAW perturbation under different conditions.

Figure 4.15 shows the standard equivalent circuit model that may be fitted to a two-port SAW device at frequencies near resonance (4). This is almost identical to the circuit for a bulk-wave thickness shear-mode device (22) with the addition of a second capacitance for the second IDT. (The TSM is a one-port device.) Here, the parameters R_m, C_m, and L_m are associated with the motional properties of the piezoelectric substrate. Frictional losses within the quartz and acoustic losses through compressional wave generation in a fluid in contact with the surface are represented by R_m. Both C_m and L_m represent the substrate elasticity and mass, respectively; L_m also includes the effects of any surface coating and viscous coupling with a fluid in contact with the surface. Clearly, these contributions can be broken down further, depending on the physical model for the physicochemical processes that perturb acoustic wave propagation. Electrical considerations, such as the parasitic capacitances associated with the electrical connections to the IDTs, may also be incorporated (51). Some discretion needs to be exercised because multiple linear regression can be used to fit any sufficiently complex circuit to the experimental impedance characteristics. The fact that such a circuit may provide a good fit, however, does not imply validation of the physical model behind it. Rather, the model must be clearly justifiable on physical grounds *before* attempting a fit to experimental data.

(a)

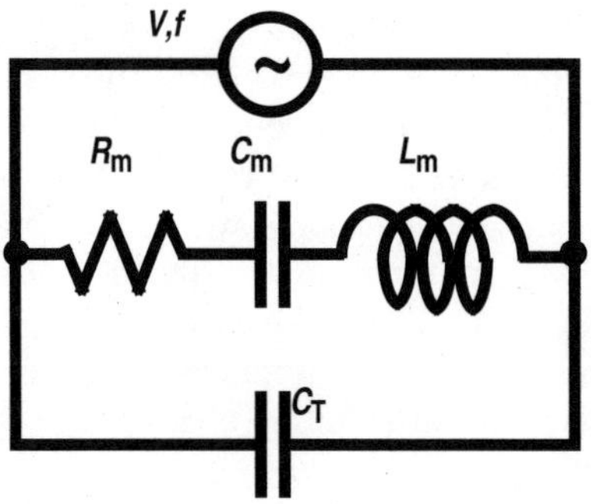

(b)

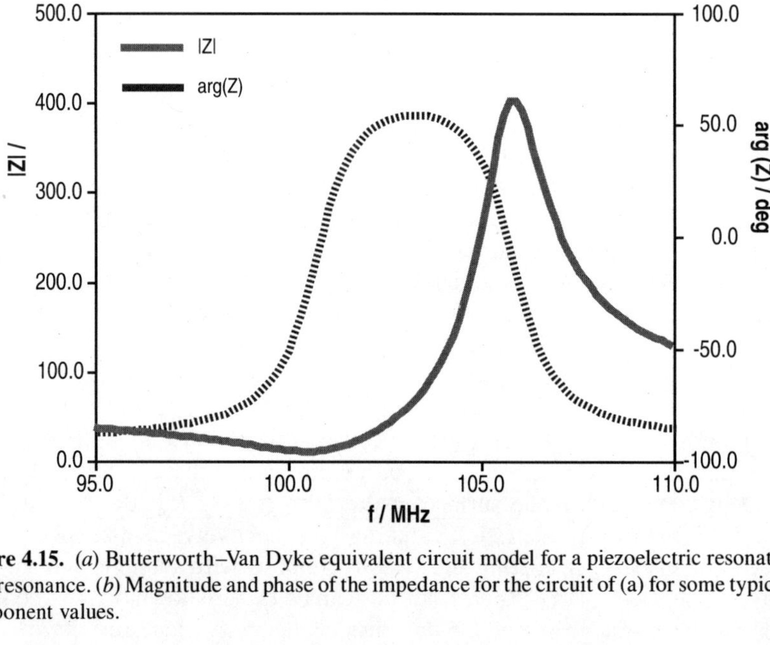

Figure 4.15. (*a*) Butterworth–Van Dyke equivalent circuit model for a piezoelectric resonator near resonance. (*b*) Magnitude and phase of the impedance for the circuit of (a) for some typical component values.

4.6. SIGNAL REFERENCING

In the preceding sections, we saw that one source of signal noise is small phase differences within the electronic circuit. Thermal drift within the electronics can also give rise to such changes, so proper heat sinking or thermostating of the electronics is important. We also saw in Section 3.2 that SAW sensor response can be strongly temperature- and pressure-dependent, especially with substrates such as $LiNbO_3$. For these reasons, many SAW chemical sensor studies have employed either two devices on the same physical substrate *or* two separate devices within the same sample chamber. One device is treated with a selective sorptive coating or polymer film, while the second is left untreated. This has an additional benefit for frequency-based measurements in that the difference frequency (of the order of kilohertz) is easier to measure and record than the absolute oscillation frequency (of the order of mega- to gigahertz).

The theory behind having two channels on one substrate is that fluctuations in temperature and pressure will have an equal effect on both channels, so that the difference in frequency (or phase) between the two will be immune to such changes. It is also often assumed that measuring the difference frequency in this manner will cancel nonspecific interactions such as surface condensation (this is discussed in more detail in Chapter 5). Experience has shown, however, that the two channels rarely show identical behavior owing to such factors as thermal and stress gradients across the substrate. Acoustic coupling or cross-talk between the sample and reference channels can be also problematical, especially for very thin films that do not produce a large frequency shift due to coating alone. It is thus not clear whether the apparent advantages of dual channel devices are ever fully realized in practice. An alternative approach being investigated by Alder and co-workers (52–54) is to derive a mathematical function describing the behavior to be compensated for, and then apply this function to "correct" the measured frequency values before analyzing the data. This approach has been applied to the effects of both temperature (52) and humidity (53,54).

We have already mentioned that acoustic coupling of adjacent devices can be avoided by using additional components to tune the individual frequencies. In the simplest case, a capacitor is connected in parallel with the receiving IDT in a delay-line oscillator. It has been found, however, that this can adversely affect sensor response behavior when very thin films are used such that the test vapors can readily penetrate between the fingers of the IDTs (16). In fact, for uncoated devices it is quite possible for the response to organic vapors to be reversed, depending on the amount of parallel external capacitance. This effect is a function of the bandwidth and amplification factor of the external circuit, along with its ability to accommodate large phase shifts.

We have also mentioned the importance of impedance matching. For the most efficient driving and monitoring of IDTs, it is necessary that the impedances of the source and generating IDT, and the detector and receiving IDT, be matched in terms of their electrical impedances over the frequency range of interest. This is again achieved by introducing electrical components in the external circuit to compensate for the capacitive–inductive nature of the IDTs. Since the impedance properties of SAW devices are predominately capacitive in nature, this may involve connecting an inductor (r.f. choke) in series with each IDT to achieve a nominal impedance of 50 Ω. This ignores the fact that the impedance properties of the IDTs are dynamic in nature, however, and will vary throughout vapor sorption–desorption cycles. In general, it has been assumed that such changes will be negligible and arguably will be when the IDTs are coated with a thick layer of a nonpolar material such that high dielectric vapors do not penetrate to the transducer surface. For polar, dielectric coatings, and uncoated devices, however, this assumption may well be unfounded and should be verified experimentally. In addition, it is possible to design impedance-matching networks (rather than employing single components) which can decouple such capacitive effects from the external circuit.

REFERENCES

1. D. P. Morgan, Ed., *Key Papers on Surface-acoustic-wave Passive Interdigital Devices,* IEE Reprint Series, Vol. 2, Peter Peregrinus Ltd., Stevenage, UK, 1976.
2. B. A. Auld, *Acoustic Fields and Waves in Solids,* Vol. 2, Wiley-Interscience, New York, 1973, pp. 170–176.
3. A. A. Oliner, Ed., *Acoustic Surface Waves,* Springer-Verlag, New York, 1978.
4. S. Datta, *Surface Acoustic Wave Devices,* Prentice-Hall, Englewood Cliffs, NJ, 1986.
5. M. E. Barnard, M. J. Lancaster, and E. G. S. Paige, *IEEE Trans. Ultrason. Ferroelec. Freq. Control,* **1989,** *UFFC-36,* 565.
6. H. R. Kaiser and P. S. Castro, *Proc. Inst. Radio Eng.,* **1962,** 2142.
7. H. Engan, *IEEE Trans. Electron. Devices,* **1969,** *ED-16,* 1014.
8. B. A. Auld and G. S. Kino, *IEEE Trans. Electron. Devices,* **1971,** *ED-18,* 898.
9. J. I. Smith, *IEEE Trans. Microwave Theory and Tech.,* **1971,** *MTT-19,* 424.
10. R. Y. Scapple, *IEEE Proc. 1974 Electron. Comp. Conf., Washington DC,* **1974,** IEEE Cat. #74CH0853-2PHP, 203.
11. L. Binotto and G. F. Piancentini, *Thin Solid Films,* **1972,** *12,* 325.
12. M. Lui, A. R. King, V. Jaccarino, and G. L. Snider, in *Growth, Characterization and Properties of Ultrathin Magnetic Films and Layers,* B. T. Jonker, J. P. Heremans, and E. M. Marinero, Eds., *Mater. Res. Soc. Symp. Proc.,* **1989,** *151,* 225.
13. K. Alberti, J. Haas, C. Plog, and F. Fetting, *Catalysis Today,* **1991,** *8,* 509.
14. D. E. Gray, Ed., *American Institute of Physics Handbook,* 2nd ed., McGraw-Hill, New York, 1963, pp. 5–15.
15. H.-E. Endres and S. Drost, *Sens. Actuators B,* **1991,** *B4,* 95.
16. D. C. Stone and M. Thompson, *Anal. Chem.,* **1993,** *65,* 352.

17. G. Arfken, *Mathematical Methods for Physicists,* 3rd ed., Academic Press, New York, 1985, pp. 321–325.
18. W. H. Press, S. A. Teukolsky, W. T. Vetterling, and B. P. Flannery, *Numerical Recipes in C: The Art of Scientific Computing,* 2nd ed., Cambridge University Press, Cambridge, UK, 1992, pp. 261–271.
19. J. Lin, S. Möller, and E. Obermeier, *Sens. Actuators B,* **1991,** *B5,* 223.
20. J. Duclos and M. Leduc, *Revue Phys. Appl.,* **1982,** *17,* 75.
21. W. R. Smith, *IEEE Trans. Microwave Theory Trans.,* **1974,** *MTT-22,* 960.
22. V. E. Bottom, *Introduction to Quartz Crystal Unit Design,* Van Nostrand Reinhold, New York, 1982, Chapter 6.
23. W. P. Mason, *Electromechanical Transducers and Wave Filters,* 2nd ed., Van Nostrand, Princeton, NJ, 1948, pp. 201–209, 399–409.
24. W. R. Smith, H. M. Gerrard, J. H. Collins, T. M. Reeder, and H. J. Shaw, *IEEE Trans. Microwave Theory Trans.,* **1969,** *MTT-17,* 856.
25. A. J. Ricco, S. J. Martin, and T. E. Zipperian, *Sens. Actuators,* **1985,** *8,* 319.
26. H. Wohltjen, *Sens. Actuators,* **1984,** *5,* 307.
27. S. J. Martin, S. Datta, R. L. Gunshor, R. F. Pierret, M. R. Melloch, and E. J. Staples, *IEEE Ultrason. Symp. (Proc.),* **1982,** 290.
28. S. J. Martin, K. S. Schweizer, A. J. Ricco, and T. E. Zipperian, *Proc. of the Int. Conf. on Solid-State Sensors and Actuators, Pennsylvania PA June 11–14th 1985,* IEEE Cat.# 85CH2127-9, **1985,** 71.
29. C. Caliendo, A. D'Amico, P. Verardi, and E. Verona, *IEEE Ultrason. Symp. (Proc.),* **1988,** 569.
30. M. S. Nieuwenhuizen, A. J. Nederlof, M. J. Vellekoop, and A. Venema, *Sens. Actuators,* **1989,** *19,* 385.
31. C. T. Chuang and R. M. White, *IEEE Ultrason. Symp. (Proc.),* **1982,** 295.
32. R. M. White, P. J. Wicher, S. W. Wenzel, and E. T. Zellers, *IEEE Trans. Ultrason. Ferrorelec. Freq. Control,* **1987,** *UFFC-34,* 162.
33. P. W. Walton, P. M. Gibney, M. P. Roe, M. J. Lang, and W. J. Andrews, *Analyst,* **1993,** *118,* 425.
34. E. A. Ash, in *Acoustic Surface Waves,* A. A. Oliner Ed., Springer-Verlag, New York, 1978, pp. 97–185.
35. T. E. Parker, *IEEE Ultrason. Symp. (Proc.),* **1982,** 268.
36. N. Rezgui and J. F. Alder, *Anal. Proc.,* **1989,** *26,* 46.
37. S.-M. Chang, B. Ebert, E. Tamiya, and I. Karube, *J. Biotechnol.*, **1990,** *16,* 211.
38. V. M. Hietala, G. C. Frye, and S. J. Martin, *Langmuir,* **1993,** *9,* 249.
39. J. G. Brace, T. S. Sanfelippo, and S. G. Joshi, *Sens. Actuators,* **1988,** *14,* 47.
40. G. C. Frye, S. J. Martin, A. J. Ricco, and C. J. Brinker, in *Chemical Sensors and Microinstrumentation,* R. W. Murray, R. E. Dessy, W. R. Heineman, J. Janata, and W. R. Seitz, Eds., *ACS Symp. Ser.,* **1989,** *403,* 208.
41. D. Hauden, G. Jaillet, and R. Coquerel, *IEEE Ultrason. Symp. (Proc.),* **1981,** 148.

42. M. Thompson and D. C. Stone, *Anal. Chem.,* **1990,** *62,* 1895.
43. S. J. Martin, A. J. Ricco, D. S. Ginley, and T. E. Zipperian, *IEEE Trans. Ultrason. Ferroelec. Freq. Control,* **1987,** *UFFC-34,* 142.
44. A. D'Amico, A. Palma, and E. Verona, *Sens. Actuators,* **1982/83,** *3,* 31.
45. B. Holcroft and G. G. Roberts, *Thin Solid Films,* **1988,** *160,* 445.
46. H. Wohltjen and R. Dessy, *Anal. Chem.,* **1979,** *51,* 1458.
47. R. S. Falconer, J. F. Vetelino, D. J. Smith, and M. J. Osborn, *IEEE Ultrason. Symp. (Proc.),* **1990,** 315.
48. S. J. Martin, G. C. Frye, and S. D. Senturia, *Anal. Chem.,* **1994,** *66,* 2201.
49. A. D'Amico, A. Petri, P. Verardi, and E. Verona, *IEEE Ultrason. Symp. (Proc.),* **1987,** 633.
50. Y. Inoue and M. Matsukawa, *J. Chem. Soc., Chem. Commun.,* **1990,** 296.
51. B. H. Horine and D. C. Malocha, *IEEE Ultrason. Symp. (Proc.),* **1990,** 477.
52. J. F. Alder, C. G. Fox, A. R. M. Przybylko, N. D. Rezgui, and R. D. Snook, *Analyst,* **1989,** *114,* 1163.
53. C. G. Fox and J. F. Alder, *Anal. Chim. Acta,* **1991,** *248,* 337.
54. F. Benmakroha and J. F. Alder, *Anal. Chim. Acta,* **1995,** *302,* 255.

CHAPTER

5

SENSOR COATINGS AND SAMPLE GENERATION

In the previous chapter we examined in some detail the electrical and practical aspects of SAW device operation. To produce viable chemical sensors, however, it is necessary to modify the surface of these devices to achieve chemical selectivity and sensitivity. This structure must then be interfaced with a sample/standard generation and handling system to achieve detection and measurement of the target analyte. In practical terms, we require a signal arising from acoustic wave perturbation that can reliably and reproducibly be related to the gas- or liquid-phase concentration of the target analyte. What we actually measure, however, is a function of the analyte concentration *at the sensor surface.* The response function depends on temperature, pressure, mass, viscosity, conductivity, dielectric constant, film thickness, and so on. The analytical utility of our measurement, therefore, is critically dependent on (1) precise control of the experimental conditions, (2) device fabrication and coating, and (3) adequate calibration. This in turn requires careful attention to the information required from the measurement, the physical realization of the complete sensor system, and a good understanding of the chemical system involved. In this chapter we consider aspects of coating selection and deposition, sample generation and presentation, and calibration. The primary focus is on gas-phase systems and measurements. Liquid-phase systems are discussed separately where appropriate.

5.1. RESPONSE OF SAW CHEMICAL SENSORS

In considering SAW sensor response to challenges of a particular analyte, it is important to realize that the processes involved are essentially kinetic in nature. Indeed, there are many significant parallels between acoustic sensor systems in general and chromatography. This is illustrated in Figure 5.1, which shows a typical scenario for a gas-phase SAW chemical-sensor measurement. The target analyte molecule must first be transported to the surface, where it may become physisorbed. In this particular example, the analyte is then shown diffusing into the bulk of a surface coating (perturbing acoustic wave propagation in the process) before returning to the surface and subsequently desorbing. It is equally valid, however, to envision the case where the

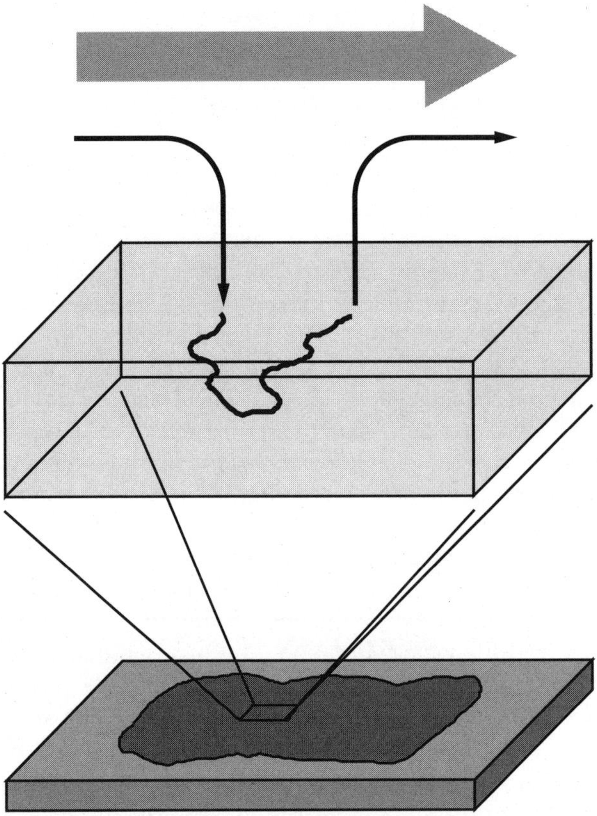

Figure 5.1. Representation of vapor-phase sorption, diffusion, and desorption with a coated chemical sensor.

target analyte simply adsorbs to the surface (also causing acoustic wave perturbation) before desorbing. In either case, there is (1) a transport process to the surface, (2) a residence time at or in the surface, and (3) a corresponding transport process from the surface. Clearly, then, we need to consider analyte diffusion properties, interaction energies, and solvation parameters, all of which show temperature dependence.

We can also identify two measurement conditions: (1) the *static* case, where exposure to the target analyte occurs for a fixed amount of analyte in a fixed volume of gas; and (2) the *dynamic* case, where the bulk (or carrier) phase is being continually transported across the surface of the sensor. We can further consider the dynamic cases where (a) the analyte concentration is constant throughout the measurement (steady state) and (b) the analyte concentration varies from zero through a maximum back to zero (transient). The same analysis applies to the liquid phase, although in this case rates of diffusion to and from the surface from the liquid phase will obviously be slower than in the gas phase.

5.1.1. Static Measurements

Static measurements are used to investigate the properties of thin films deposited onto the SAW device. Some specific examples are the measurement of the diffusion properties of very thin polymeric fims and surface area and porosity measurements on sol-gel fims (1,2). The advantage of SAW devices for these measurements lies in their sensitivity to very small amounts of material. More conventional methods of performing these measurements require bulk samples of the materials, but bulk and thin-film properties can be markedly different. Interpretation of thin-film diffusion data is complicated by the range of mechanisms that can contribute to observed frequency, phase and attenuation changes (Chapter 3), and the lack of relevant physical data for many polymeric materials at the frequencies typically employed.

The SAW devices may be incorporated into existing instrumentation for measurements on small samples or custom-test apparatus may be constructed. Key considerations are precise control of temperature and sample vapor partial pressure. Care must also be taken to prevent r.f. interference by adequately shielding the apparatus in an earthed, metal enclosure or Faraday cage. Ideally, the electronics for the SAW measurement should be connected using very short cables (Section 4.5.1).

5.1.2. Dynamic Measurements

The majority of published SAW chemical sensor applications fall into this category. Samples and standards are presented to the SAW device in a sequential fashion. Measurements are usually made on the steady-state signal, although derivative measurements have also been performed. Control of temperature and pressure are again important, as are r.f. shielding and electrical connections. Calibration standards are usually obtained by dilution of the sample gas stream with pure carrier gas. Stable flow rates and correct calibration of gas-phase vapor concentrations are therefore essential. Liquid-phase samples may be handled (for appropriate devices) using techniques developed in liquid chromatography and flow-injection analysis. Transient measurements may also be performed using a gas-chromatographic configuration in which the SAW device functions as the gas-chromatographic detector (3–5). We discuss the relative merits of such systems in Section 5.4.

5.1.3. Sensor Response Theory

Considering Figure 5.1 in more detail, we can see that as the target analyte reaches the sensor, it must first adsorb to the sensor surface. This process can occur either through weak physical interactions ($\Delta H < 40$ kJ mol^{-1} of adsorbate) or chemical bond formation ($\Delta H \approx 80 - 400$ kJ mol^{-1} of adsorbate) (6).

Heats of physical adsorption ("physisorption") are of the same order of magnitude as heats of liquefaction—that is, physical adsorption is similar to condensation. Physisorption will therefore occur to an appreciable extent under conditions that favor liquefaction. Physisorption is characterized by van der Waals and hydrogen bonding interactions, which are readily reversible. The initial adsorption can be followed by mulitlayer buildup, resulting in a film that can behave as a two-dimensional liquid.

Chemical adsorption ("chemisorption") usually results in monolayers, occurs more readily at lower temperatures and higher pressures than physisorption, and is less easily reversed. Some examples include the dissociative adsorption processes found in metal surface oxide and hydride film formation, and the strong binding of water to such surfaces. Other possible interactions that lie between chemisorption and physisorption (but are grouped with physisorption for our purposes) are Brønsted–Lowry and Lewis acid–base interactions. In the case of highly nonpolar analytes and sensor coatings, we can also include hydrophobic effects, which are essentially driven by entropic considerations (7).

Considering only physisorption, it is clear that temperature will have a strong influence on sensor response by altering the rates of surface adsorption and desorption. Other factors affecting sensor response are the available surface area for adsorption and the surface density of active sites. Therefore, low surface temperatures and porous films give increased chemical sensitivity and better limits of detection, at the expense of increased response and recovery times for either static or dynamic measurements. Another important consideration here is sensor selectivity, because at low temperatures extensive condensation effects may be anticipated. An analyte showing a strong affinity for a given surface will more readily condense on that surface because of the "nucleation" effect of physisorption. Vapor molecules with no (or little) affinity for the surface will condense when conditions of temperature and concentration are such that droplet formation occurs in the gas phase with subsequent deposition onto all available surfaces. To achieve selectivity, then, requires increased surface temperatures (to prevent condensation effects) and incorporation of functionalities that promote physisorption with (and only with) the target analyte.

The chemical and phsyical nature of the sensor surface (coating) is reflected in the adsorption isotherm for the target analyte. The simplest case for

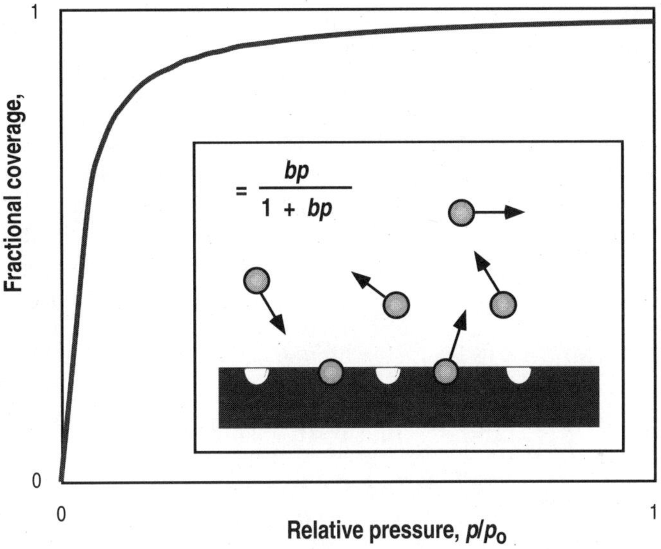

Figure 5.2. Langmuir isotherm for monolayer adsorption: fractional coverage as a function of relative vapor pressure. Inset: model assumed in deriving the Langmuir isotherm.

gas adsorption at a surface is the Langmuir isotherm. This is a kinetic derivation based on an ideal localized monolayer (Fig. 5.2). It assumes that molecules striking the surface are adsorbed at definite points, or "sites"; that only one molecule can occupy a site at a time; and that there is no lateral interaction between molecules adsorbed on neighboring sites:

$$\theta = \frac{bp}{1 + bp} \tag{5.1}$$

where θ is the fraction of occupied sites at equilibrium for a given partial pressure p. The constant b accounts for the relative rates of adsorption and desorption and the fraction of molecules striking the surface that are captured by it. Various refinements of the original isotherm are available; however, all suffer from the implicit assumption of surface homogeneity (all sites show the same "stickiness" toward the adsorbate) and the exclusion of lateral interactions (6). Similarly, the b term can be derived from the kinetic theory of gases or defined explicitly by a thermodynamic approach:

$$b = \frac{h^3}{kT(2\pi mkT)^{3/2}} \frac{f_a(T)}{f_g(T)} \exp\left(\frac{q}{kT}\right) \tag{5.2}$$

where $f_a(T)$ and $f_g(T)$ are the internal partition functions for a molecule in the adsorbed and gaseous states, respectively, q is the energy required to move a molecule from the lowest adsorbed to the lowest gaseous state, and h is Planck's constant.

Another problem is that, for physisorption, multilayer buildup is quite common. The result can be exponential adsorption isotherms which may require initial Langmuirian adsorption, and which may be limited to a maximum of n layers at the saturation vapor pressure (p_0). Such isotherms have been classified by Brunauer and co-workers as types I–V (8,9), and may be described mathematically through the BET or n-layer BET isotherms (6) (Fig. 5.3):

$$V^s = \frac{V_m^s Cp}{(p_0 - p)\{1 + (C - 1)p/p_0\}} \tag{5.3}$$

$$V^s = \frac{V_m^s Cx}{(1 - x)} \frac{[1 - (n + 1)x^n + nx^{n+1}]}{[1 + (C - 1)x - x^{n+1}]} \tag{5.4}$$

with

$$C = \exp\left(\frac{E_1 - E_L}{RT}\right)$$

where V^s is the volume adsorbed, V_m^s is the volume of a complete ideal localized monolayer, $x = p/p_0$, n is the number of layers, E_1 is the first layer heat of adsorption, and E_L is the heat of liquefaction of the bulk liquid.

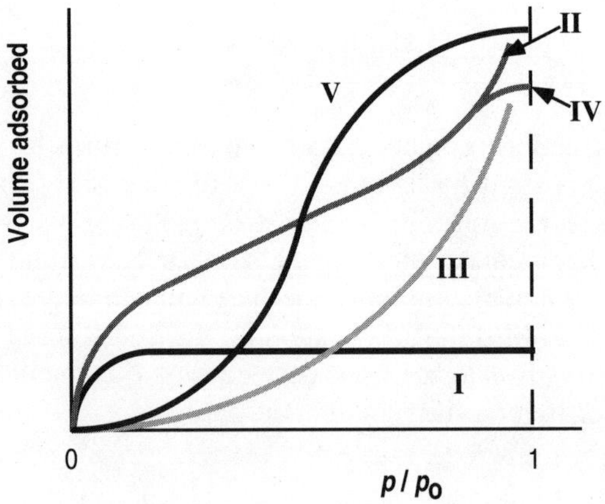

Figure 5.3. Shapes of the different adsorption isotherms as classified by Brunauer et al. (8,9).

Such isotherms can allow estimation of specific surface areas and porosities, the latter being reflected by hysteresis between the ascending and descending isotherms.

For rigid, porous films such as sol-gels, sensor response may be described entirely as a result of the physisorption process and the sensor response described using the equations developed by Martin et al. (Section 3.6.1). For polymeric materials, sensor response now becomes a function of the solubility of the analyte in the coating. For a sample that is completely insoluble in the polymer, the response remains due to purely surface phenomena, but for other samples absorption becomes an important factor. It is this class of coated SAW sensors that most obviously invite a comparison with the use of stationary-phase coatings in GLC. In fact, the first SAW sensor coatings for vapor detection were stationary phase materials (3). By analogy, then, it was initially predicted that SAW sensors could be used to obtain physical and thermodynamic information about polymers. While the situation is considerably more complex than originally believed (Section 3.7), it remains true that SAW chemical sensors can be used to examine complex adsorption and solubility properties for a wide range of materials.

5.1.4. Implications of Sensor Response Theory

We have seen that SAW sensor response involves several kinetic phenomena including adsorption, diffusion, and desorption. Further, the rates of these processes (and corresponding dynamic equilibrium response values) will be dependent on both temperature, film properties (thickness, porosity, etc.) and analyte concentration. In other words, an instantaneous vapor- or liquid-phase analyte concentration, C_v, gives rise to a corresponding sensor-phase analyte concentration, C_s, generating the resulting sensor response. In general, we can expect C_s to be a function of, but not equal to, C_v. Further, unless the sorption process is essentially irreversible under the conditions used, $C_v < C_s$.

For coated chemical sensors, it therefore becomes important to normalize sensor response to take into account variations in coating thickness and porosity, especially between different devices. Different approaches to this issue will be described below in the context of coating selection. Another implication is that careful characterization, calibration, and control of the sensor system are necessary, especially if mechanistic or thermodynamic studies are to be based on the results. The importance of this issue cannot be underestimated. We can also apply our response theory to the prediction of dynamic response profiles to challenges of a target analyte under controlled conditions.

If we first consider a coated SAW sensor responding to challenges of a sample vapor which shows Langmuirian adsorption, the expected dynamic steady-state response curve for a step concentration function (assuming a pure mass-loading mechanism) would look like that of Figure 5.4*a*. This comprises two regimes: (1) a fast initial response due to surface adsorption, up to monolayer coverage; and (2) a slower, diffusion-controlled response due to absorption within the bulk of the polymer. Within this second regime, we would expect that the frequency change would vary with the square root of elapsed time (1). If at time $t = 0$ the polymer film is in equilibrium with a sorbate partial pressure P_1 to give a film concentration $C_0(P_1)$, the total moles absorbed by the film as a function of time is shown to be

$$M(t) = M_{\max}\left(1 - 2\sum_{n=1}^{\infty} \frac{\exp(-\psi^2 Dt/h^2)}{\psi^2}\right) \tag{5.5}$$

where $M_{\max}$ is the total moles absorbed at final equilibrium, $\psi = \pi(n - 0.5)$, D is the diffusion coefficient for the sorbate in the film, and h is film thickness.

When the input concentration is returned to zero, the reverse process occurs. The effect of temperature in this case is readily anticipated. Increasing the temperature will speed up the rates of diffusion through the bulk of the polymer and desorption, resulting in (1) a lower concentration in the polymer phase at the steady state equilibrium and (2) faster rise and fall curves on the response profile. On the other hand, lowering the temperature will result in (1) a higher concentration in the polymer phase at equilibrium and (2) slower rise and fall curves. In other words, sensor response time may be reduced by increasing device temperature, but only at the expense of sensitivity and limit of detection.

The response profile shown in Figure 5.4*a* is "symmetric" in the sense that the rise and fall curves are identical in shape, the fall curve simply being the inverse of the rise curve. Other possibilities may occur, however, if different mechanisms are in operation. The most obvious case would be chemisorption, for which the response would be irreversible and have a finite limit (Fig. 5.4*b*). Another example would be if the polymer film of Fig. 5.4*a* was porous, resulting in a hysteresis in the adsorption isotherm. In this case, the fall curve would not be the inverse of the rise curve because desorption would be slower.

Perturbation of the polymer properties by increasing vapor sorption will also have an influence, especially if the device is operating under conditions for which film viscoelastic effects are significant. Under these circumstances, we can expect to see a two-phase behavior in which the initial portion of the response curve reflects mass loading due to vapor sorption and the later

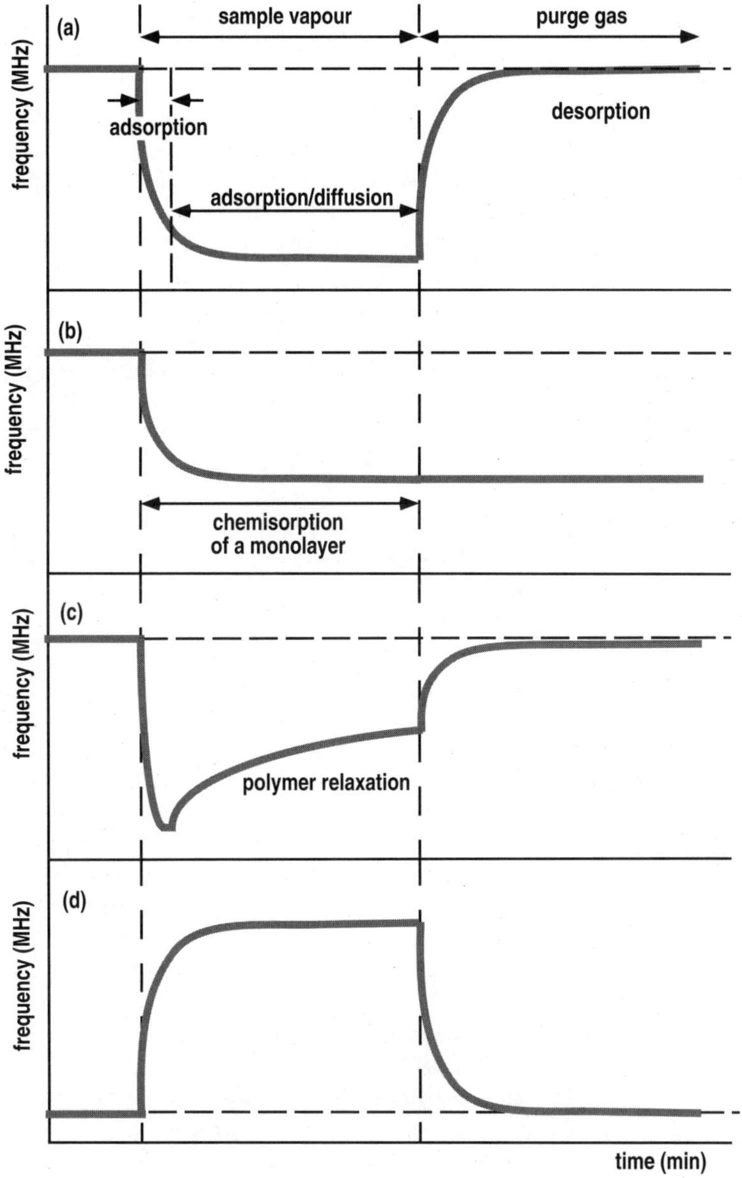

Figure 5.4. Possible response profiles of a coated SAW chemical sensor to sample vapor. (*a*) Ideal response; (*b*) irreversible chemisorption; (*c*) combined mass and viscoelastic effects; (*d*) large change in polymer properties for a small mass absorbed.

section reflects increasing changes in polymer properties with increasing time of vapor sorption. Because viscoelastic effects can result in both positive and negative frequency perturbations, it is quite conceivable that either of the responses shown in Figures 5.4*c* and *d* may result. In fact, profiles such as these have already been reported in the literature (1,10).

From our phenomenological description of sensor response, then, we can see the importance of correctly understanding the nature of sample-coating interactions in designing and evaluating chemical sensors. In particular, the nature of the adsorption isotherm *at a given temperature* can have a significant role in determining analytical sensitivity and limit of detection. Further, combining such information with the concentration profile presented to the device allows either prediction of sensor response *or* interpretation of the response to obtain physical chemical parameters for the sample-coating system under study.

5.2. RESPONSE NORMALIZATION AND INTERACTION MECHANISMS

Of particular interest in SAW chemical sensor development is the nature of specific analyte-coating interactions, and the development of methods for rational coating selection for improved sensitivity and selectivity. The first steps in this direction were the correlation of sensor response with solubility properties by pattern recognition (11) and the application of partition coefficients (12). Semiempirical molecular orbital calculations have also been used to calculate enthalpies for hydrogen-bond interactions in an attempt to correlate structural features with sensor-response data (13).

In all three cases, measurements were performed using dual-delay line oscillators, where the frequency of a delay line treated with some chemically selective material was measured relative to that of an uncoated delay line (Fig. 5.5). The resulting change in the difference frequency, ΔF, was assumed to be linearly related to the amount of analyte sorbed by the sensor coating *over and above* the amount nonspecifically adsorbed to the surface as measured by the uncoated delay line. This is not unreasonable, especially if the film has a high specific-surface area or readily solvates the analyte molecules. In the first two cases, coatings were polymeric films air-brushed onto the delay line until a certain frequency shift was observed. Exact film thicknesses were not calculated due to a lack of the necessary data; however, for one film (12) this was estimated to be $\approx$40 nm using Equation 3.5. In the third case, the coating was formed by direct reaction of the quartz surface with (3-aminopropyl) triethoxysilane (APTES), and was estimated to be 6.5 nm (7 layers) thick (13).

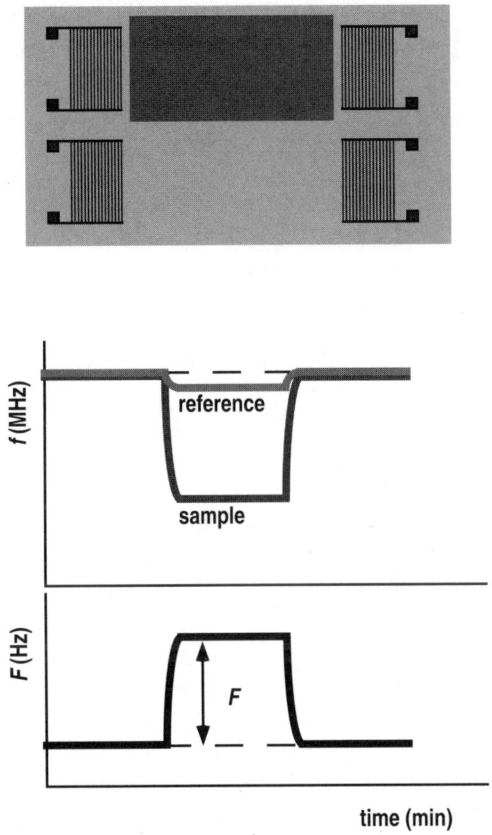

Figure 5.5. Sample referencing scheme typically employed with SAW chemical sensors to reduce temperature and nonspecific adsorption effects.

The results of these studies are interesting for a number of reasons, the principal one being that there is a strong correlation between normalized sensor response and the molecular properties of both analyte and coating material. All three also illustrate the importance of response normalization.

5.2.1. Sensor Partition Coefficients

The distribution of the analyte between the vapor and sorbent phases is characterized by the sensor partition coefficient:

$$K = C_s/C_v \tag{5.6}$$

By assuming that the frequency perturbation due to film deposition and vapor sorption are linearly related to the *mass loading,* Grate et al. (12) derived the following equation for calculating K from SAW sensor responses:

$$K_{\text{SAW}} = \frac{\Delta f_{\text{v}} \rho}{\Delta f_{\text{s}} C_{\text{v}}} \tag{5.7}$$

where Δf_{s}(kHz) and Δf_{v}(Hz) are the frequency shifts induced by the film coating and vapor sorption, respectively, ρ is the film density (g mL^{-1}), and C_{v} is in g L^{-1}. Note that if sensor response cannot be approximated by pure mass loading, K_{SAW} will not be the same as the value obtained by GLC methods, and may vary with vapor concentration. Such effects are likely to occur for high vapor concentrations and thick films. Temperature is also an issue, because solvation and diffusion processes are both affected—K values are known to be temperature-dependent. The issues of temperature and analyte solvation effects on sensor response as they affect Equation 5.7 have been addressed as described in Section 3.7.3 (15).

One interesting feature of the K_{SAW} concept is the way it explains sensor sensitivity and selectivity. For example, in the original work of Grate et al. (12), log K_{SAW} values are given that lie between 2.0 and 6.3. Assuming conditions for which Equation 5.7 is valid, this indicates that C_{s} is in excess of C_{v} by a factor of 100 to 2×10^6. This in effect represents a "preconcentration" of the analyte by the film, leading to dramatic improvements in sensitivity over an uncoated device. These results were obtained using a fluoropolyether with hydroxy side chains as the coating, which is expected to show strong hydrogen bonding interactions with appropriate vapor-phase species. This is confirmed by (1) the magnitude of the log K_{SAW} values for polar species capable of hydrogen bond interactions and (2) the significantly greater concentrations of nonpolar species required to attain K_{SAW} values that were four orders of magnitude lower. For example, the log K_{SAW} values cited above were obtained for $\approx$100 mg L^{-1} isooctane but only $\approx$50 μg L^{-1} *N,N*-dimethylacetamide, respectively. Use of a suitable coating thus promotes selectivity between analyte and interferents in the same way that choice of an appropriate stationary phase allows separation of sample components in chromatography, and for the same reasons.

At this point, it should be noted that there are two divergent approaches to the design and selection of selective coatings, especially for structurally and functionally similar analytes. One is the isolation or synthesis of highly selective "receptors" for the target analyte (e.g., through antibodies or biomimetic systems); the other is the use of an array of sensors coated with materials showing different sensitivities to the sample components combined with chemometric techniques. This approach is described in more detail in Chapter 6.

The K_{SAW} concept is clearly a useful one, although situations can arise for which it is not readily applicable. One example is where the coating density cannot be determined, and/or the sensor response to coating deposition is neither linear nor attributable to pure mass loading. An alternative normalization scheme is therefore required. Ballantine et al. (11) used calibration data to calculate the sensitivity toward each vapor as a function of coating thickness (expressed as Δf_s in kHz). This is equivalent to Equation 5.7 without the coating density in the numerator, and as such is also limited to coatings for which there is a linear relationship between coating thickness (or mass) and frequency. Heckl et al. (13) used a different approach since transient, rather than steady-state, dynamic measurements were employed. A "relative binding affinity" was calculated from the resulting peak area (Hz s) divided by the molecular weight of the vapor species. The rationale behind this was that peak area is a function of the total mass of sample injected passing over the surface in a specific volume of carrier gas (i.e., an average C_v value), and the interest was in the extent of interaction per surface sorption site (i.e., on a molecular rather than a mass basis).

It is interesting to note that both approaches were used in attempts to correlate sensor response with the type and extent of molecular physisorption interactions such as hydrogen bonding and van der Waals interactions. In the case of Ballantine et al., this was done through a comparison of the relative percentage of hydrogen bond donating–accepting groups within the different polymers and the magnitude of empirically derived solvatochromic parameters for the analyte species. Heckl et al. employed molecular modelling techniques to calculate enthalpies for hydrogen bond formation between the analyte species and the surface functional groups. These very different methods are consequences of the divergent approach to achieving sensor selectivity alluded to above. The results of both groups, however, confirm that the physisorptive mechanisms are of prime importance in sensor coating selection.

5.2.2. Linear Solvation Energy Relationships

As mentioned above, a key aim of modelling SAW chemical sensor response is the development of tools for *predicting* the response for given coating-analyte combinations in order to choose the optimum coating material for a given analyte *without* the need for extensive experimental measurements. We have also seen that the chromatographic concept of a partition coefficient has been successfully applied to chemical sensor response. It is therefore logical to extend this process to other methods developed originally for stationary phase selection in gas and liquid chromatography. Such methods include models based on analyte volatility (for gas-phase samples) and solubility in the coating (or stationary phase) material (11,12). Patrash and Zellers (16), for example, considered models based on boiling points and Hildebrand solubility parameters for gas-phase, coated SAW sensors. They conclude that the latter model is preferred when the necessary solubility data is available, although adequate approximations can be obtained from simple boiling-point data. These authors also examined the possibility of using *linear solvation energy relationships* (LSERs), which had originally been developed for gas chromatography and had also been applied to chemical sensors (12,14).

A linear solvation energy relationship attempts to relate experimental log K values to a set of parameters for any given solute that quantify its solubility properties in a given material. Multiple linear regression analysis is performed on this data, the coefficients extracted in this fashion being the corresponding solvent property parameters such that (12,14)

$$\log K = c + rR_2 + s\pi_2^* + a\alpha_2^{\mathrm{H}} + b\beta_2^{\mathrm{H}} + l \log L^{16} \tag{5.8}$$

Here c is a regression constant and the subscript 2 indicates solute parameters. Hydrogen bond acid and base strength is measured by the parameters α_2^{H} and β_2^{H}, respectively. These are derived from the equilibrium constants for complexation of the solute with a reference base or acid, respectively, in an inert solvent. The term log L^{16} accounts for dispersion interactions, and includes cavity effects. Finally, R_2 accounts for solute polarizability while π_2^* is a measure of dipole–dipole and dipole-induced dipole interactions. By obtaining both solute and solvent parameters through the appropriate experiments, tables of values can be constructed that can then be used to predict the resulting log K values for previously untried combinations. A detailed evaluation of this method for polymer-coated acoustic wave sensors was recently published by Grate et al. (17), while an extension of the concept to arrays of sensors using a principal components methodology has been described by Zellers et al. (18).

5.3. COATING MATERIALS AND METHODS

Chemical sensor coatings fulfill two complementary roles: they provide a mechanism for attaining selectivity toward the target analyte and a mechanism for improving device sensitivity. Thus an uncoated piezoelectric device could be used to monitor hydrogen pressure in the absence of any interferents. To improve sensitivity to very low levels of hydrogen, however, or enhance the preferential sorption of hydrogen over other gases, one could use a platinum or palladium film. The required goals of improved sensitivity and selectivity are met in this example by the increased capacity for, and residence time of, molecular hydrogen at the device surface vs. the reduced residence time of competing species. This example also illustrates the use of a sensor coating to provide an additional mechanism for chemical transduction, in this case through perturbation of surface conductivity (Section 3.5).

In terms of coating materials, by far the majority employed in SAW chemical sensor applications are polymeric materials chosen primarily on the basis of analyte solubility. Metal and metal oxide films have also been used, largely on the basis of previously known sorption interactions with small gas molecules such as hydrogen and nitrogen dioxide. A number of phthalocyanines and phthalocyanine derivatives have also been employed, again based on the known interaction with nitrogen dioxide. Other examples include $PtCl_2$(olefin)(amine) complexes for styrene and poly(ethylene maleate) for cyclopentadiene. Cage structures have not been exploited extensively for SAW sensors, although zeolite, cyclophane, and buckminsterfullerene (C60) coatings have all been reported. Biosensor coatings are rarely used with SAW devices, largely because of the constraints imposed by Rayleigh wave attenuation in liquids and the fact that the alternative structures described in Chapter 4 are not yet commercially available.

The choice of coating method will be dictated largely by the coating material. The principal methods are solvent evaporation, vapor deposition, organized self-assembly, physical entrapment, and covalent binding (Table 5.1). These will each be considered in more detail in the following sections. Applications of the different coatings are discussed in Chapter 6.

TABLE 5.1. Coating Type and Deposition Method

Coating	Method	Application
Antibodies, enzymes, nucleic acids, and so on	Physical entrapment or covalent binding	Biosensors, CW and BW agents, pesticides
Cyclophanes, zeolites, and so on	Physical entrapment	Low-molecular-weight gases and vapors
Lipids, phthalocyanines, and so on	Langmuir–Blodgett deposition	Organic vapors, odourants, and so on
Metals, metal oxides	Thermal or plasma deposition	Low-molecular-weight gases
Polymeric film	Spray or spin coating, solvent evaporation	Organic vapors, film properties
Thiols, silanes	Self-assembley	Organic vapors, odorants, and so on

5.3.1. Solvent Evaporation

Solvent evaporation techniques are commonly employed with polymeric materials, and include casting, painting, dip coating, and spin coating, although the method of choice is spray coating using an airbrush. This latter technique can be used in conjunction with a mask to define the chemically selective area of the device, which is especially important in the case of conducting or semiconducting films. Dilute solutions (typically <0.5% by mass) of the coating are prepared in volatile solvents such as methanol, chloroform, acetone, or toluene and coated onto the device surface. Typically, the device frequency is monitored throughout this process, and coating is continued until a specific frequency shift (Δf_s) has been obtained. If the density of the coating material is known, the film thickness can be estimated from the frequency shift making the assumptions outlined previously (Section 5.2).

It is essential to clean the device surface first to ensure a good quality film in intimate contact with the substrate. Appropriate techniques include sonicating in chloroform:methanol, washing with strong acid, or plasma cleaning. The desired result is that the coating solution wet the surface in order to obtain a continuous, uniform film. The importance of proper surface preparation prior to coating has been underlined by a recent report (19).

The coating method may well leave specific morphological features in the film. For example, studies of air-brushed polymeric coatings by Nomarski reflectance optical microscopy reveal that the resulting films tend to aggregate, probably as a result of the polymer "pulling together" as the solvent evaporates. Such features do not appear to adversely affect sensor performance unless surface conductivity mechanisms are being exploited (20,21). Spin coating can also have morphological consequences. The required coating material is dissolved in a suitable solvent, which is then spun laterally at speeds up to 4000 rpm to spread the coating across the device surface. The success of this technique depends on the viscosity of the coating solution and the volatility of the solvent. For example, Bartley and Dominguez (22) found that solutions of polycarbonate in chloroform resulted in radial "waves" in the final coating. This was avoided by using a more viscous solvent (chlorobenzene) instead. A summary of the polymeric materials used as SAW coatings is given in Table 5.2.

5.3.2. Langmuir–Blodgett Films

Langmuir–Blodgett (LB) film technology allows the deposition of molecular structures onto flat substrates. It exploits the ordering of molecules such as lipids at an air–liquid interface (35), and has been widely used for producing

TABLE 5.2. Examples of Polymeric Coatings on SAW Devices

Film	Application	Ref.
Apiezon L	Organic vapors	18
Cellulose acetate butyrate	Humidity	23
Diethylene glycol adipate	Organic vapors	18
Ethyl cellulose	Solvents	24
Fluoropolyol	CW simulants, humidity, properties, vapors	11, 15, 17, 24–27
Fomblin-ZDOL	Organic vapors	17
Hexafluoro-2-propanol polysiloxane	Organic vapors	17
N-Ethyl *o*,*p*-toluenesulphonamide	Organic vapors	18
OV-25, OV-202, OV-210, OV-215, OV-275	CW simulants, organic vapors	11, 16-18
Poly(2-hydroxyethyl acrylate)	Organic vapors	18
Poly(4-vinyl hexafluorocumyl alcohol)	Organic vapors	17
Poly(acrylate)	CW simulants	27
Poly(acrylonitrile butadiene)	CW simulants, vapors	11, 18
Poly(amidoxime)	CW simulants, vapors	11, 28
Poly(carbonate)	Properties, toluene	22, 29
Poly(chloroprene)	Organic vapors	18
Poly(diphenoxy phosphazene)	Organic vapors	18
Poly(epichlorohydrin)	CW simulants, properties, vapours	11, 15, 17, 18
Poly(ethylene)	Properties	29
Poly(ethylene maleate)	CW simulants, cyclopentadiene, properties, vapors	11, 65, 30, 31 31
Poly(ethyleneimine)	CO_2, CW simulants, humidity, vapors	24, 17, 32
Poly(ethylene terephthalate)	Properties, solvent	29
Poly(fluoroalcohol acrylate)	CW simulants	27
Poly(hexafluorodimethyl carbinolstyrene)	CW simulants	27
Poly(hexylacrylate)	Organic vapors	18
Poly(isobutylene)	organic vapors, properties	15-18, 20
Poly(isoprene)	CW simulants, vapors, properties	11, 30
Poly(methylmethacrylate)	Properties, vapors	18, 29, 30

TABLE 5.2. Continued

Film	Application	Ref.
Poly(phenyl ether)	Organic vapors	16, 18
Poly(propylene)	Properties	29
Poly(styrene)	CW simulants, properties, vapors	1, 27, 29
Poly(sulphone)	Humidity, properties	29, 33
Poly (*t*-butyl acrylate)	Film properties	20
Poly(tetrafluoroethylene)	Film properties	29
Poly(vinyl acetate)	Film properties	20
Poly(vinyl proprionate)	Organic vapors	17
Poly(vinyl tetradecanal)	Properties, vapors	20, 34
Poly(vinyl pyrrolidone)	CW simulants, humidity, properties, vapors	11, 26
Poly bis(cyanopropyl) siloxane	Organic vapors	17
Polyimide	Humidity, methanol, properties, vapors	1, 22, 23
Polyphosphazine	CW simulants, properties, vapors	11
Polysiloxane	Properties, vapors	1

organized thin films on a variety of substrates. Materials capable of forming monolayer films are characterized by having distinct hydrophilic and hydrophobic regions referred to as the headgroup and tail, respectively. The classic example of such a molecule is stearic acid [$CH_3(CH_2)_{16}CO_2H$]. Typically, a small quantity of the material is dissolved in a nonpolar solvent (such as hexane or chloroform) and allowed to spread across a water surface (the "subphase"). Solvent evaporation results in the formation of a monolayer in which the headgroups are immersed in the aqueous phase and the tails orientated away from it. A moving barrier can then be used to compress the film until a close-packed ordering of the molecules is obtained. Surface pressure–area isotherms recorded during this process reveal distinct regions associated with changes from an expanded to a compressed state. Beyond this point, further compression of the film results in collapse of the monolayer (Fig. 5.6).

Transfer of the film to a substrate is achieved by passing the latter perpendicularly through the substrate. The barrier is normally moved simultaneously during transfer to ensure constant film pressure. Transfer is usually

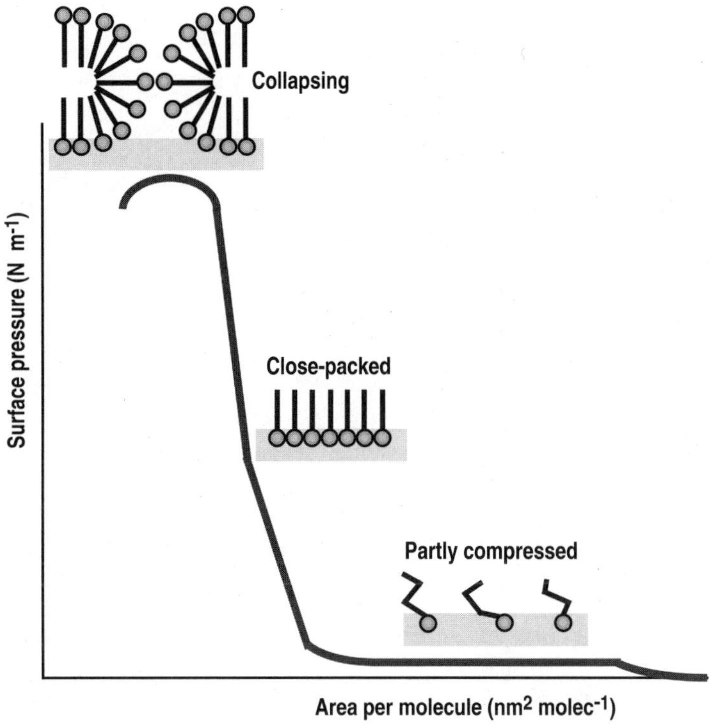

Figure 5.6. Film formation and collapse from compression of a monolayer spread on a Langmuir–Blodgett trough.

performed at surface pressures between 10 and 40 mN m^{-1}, because values outside this range may result in transfer failure or film collapse. The efficiency of the process is then measured by the transfer ratio, defined as the monolayer area removed from the water surface divided by the area of the coated substrate. Multilayer structures are built up by successive passes through the interface in either X-, Y-, or Z-type deposition (Fig. 5.7). Note that this does not guarantee that the final structure will be as indicated, however, as the layers may collapse or undergo reorientation as entrained water evaporates from the transferred film. Mixed monolayers can also be employed, in which transfer of the desired material is aided by the addition of a second material which promotes film formation and deposition. The general classes of compounds capable of forming LB films are summarized in Figure 5.8, and include fatty acids, phospholipids, porphyrins,

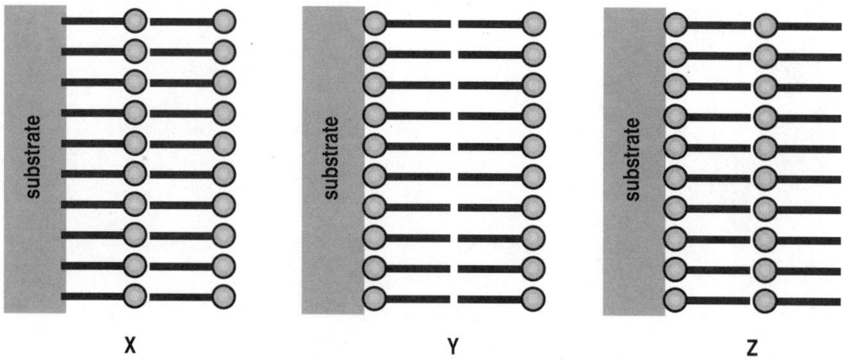

Figure 5.7. Idealized depiction of X-, Y-, and Z-type LB film-deposition methods.

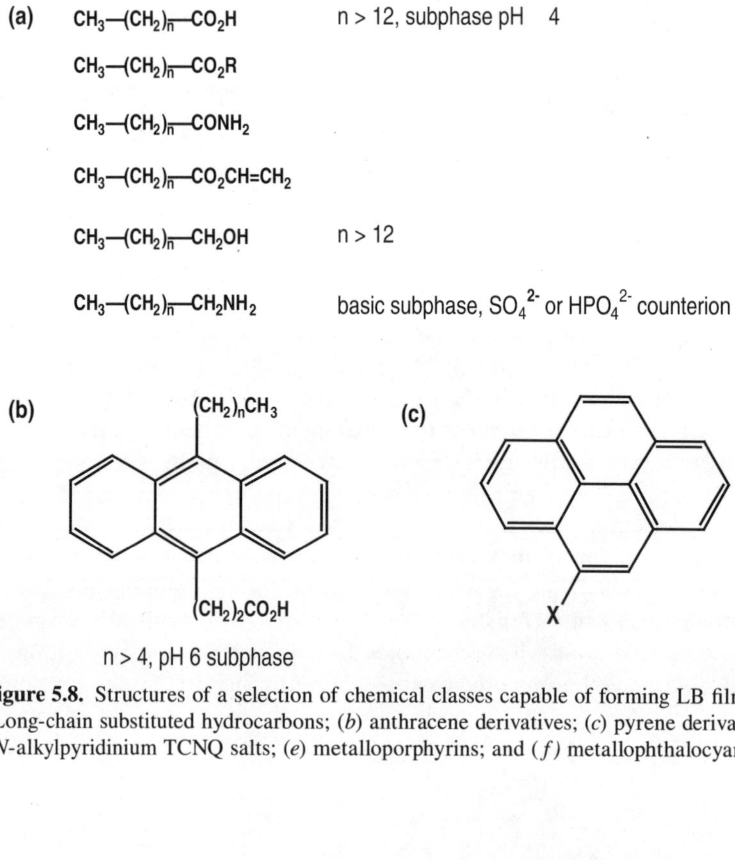

Figure 5.8. Structures of a selection of chemical classes capable of forming LB films. (*a*) Long-chain substituted hydrocarbons; (*b*) anthracene derivatives; (*c*) pyrene derivatives; (*d*) *N*-alkylpyridinium TCNQ salts; (*e*) metalloporphyrins; and (*f*) metallophthalocyanines.

(d)

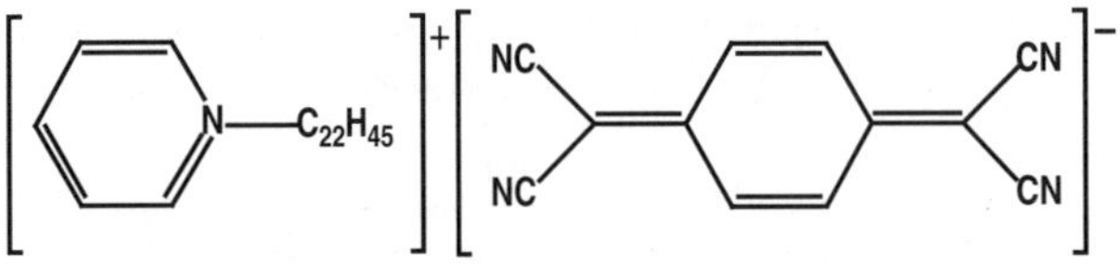

***N*-docosylpyridinium tetracyanoquinodimethane salt**

(e)

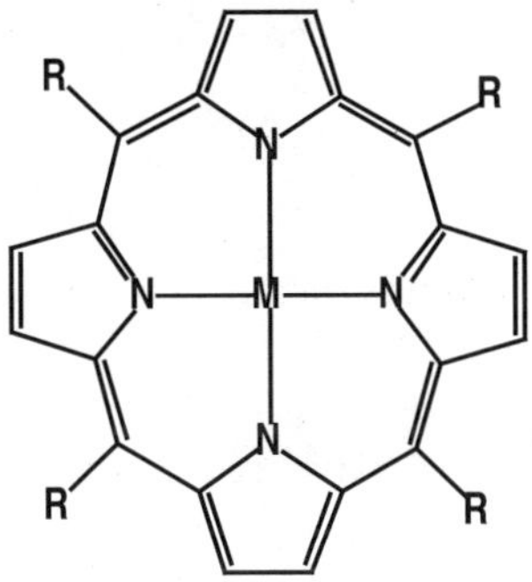

(f)

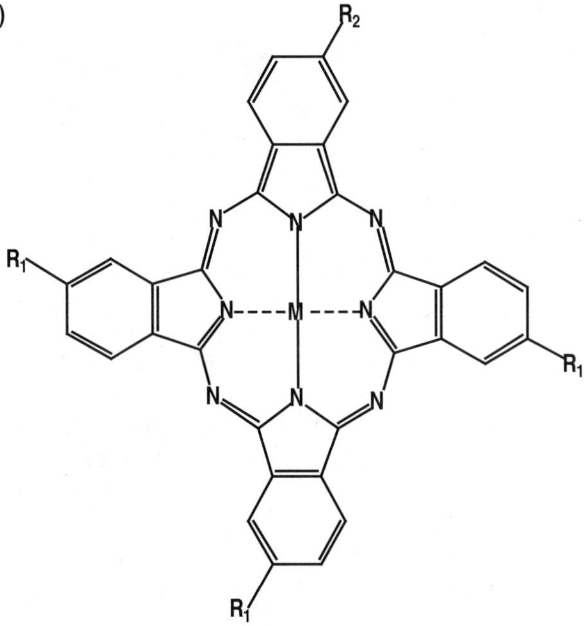

Figure 5.8. Continued

TABLE 5.3. Examples of Langmuir–Blodgett Film Deposition on SAW Devices

Film	Application	Ref.
Fluoropolyol	Dimethylmethyl-phosphonate (CW agents)	25
Metal phthalocyanines	NO_2, conductivity	21, 36, 37
Phosphatidylcholine	Odor components	38
Phosphatidylethanolamine	Odor components	38, 39
Poly(vinyl tetradecanal)	Film properties	34
Stearic acid		38
Tetracyanoquinodimethane	NO_2	40
Tetra-4-*t*-butyl silicon phthalocyanine dichloride	NO_2	41
ω-Tricosenoic acid	Acetic acid, mass	41

phthalocyanines, dyes, polycyclic aromatic compounds, and various polymers. Examples from all categories may be found in the SAW sensor literature (Table 5.3).

5.3.3. Self-Assembled Monolayers

There are many approaches to producing organized thin films of organic material on surfaces, including the use of self-assembled monolayers (SAMs) (42,43). Examples of SAMs include fatty acids, long-chain alkyl thiols and disulphides, and reactive silanes. While monolayers of many of these materials can be deposited using LB film technology, the key difference is that they form spontaneously at a solid–liquid interface through a combination of specific headgroup–surface reactions and chain–chain packing interactions. The resulting monolayers may be used as the sensor coating itself, or as the basis for subsequent surface reactions.

Self-assembled monolayers have found application in a highly diverse number of areas, including use as chromatographic and solid-phase extraction phases, modified surfaces for improved biocompatibility, wetting and adhesion properties, composite materials, and in the production of nonlinear optical devices. Although SAMs have not yet been extensively employed as SAW device coatings, they probably offer the greatest prospects for the controlled and reproducible production of chemically and physically stable selective films. For this reason, we consider the use of SAMs in more detail than LB films or polymer coatings.

The use of reactive silanes to produce SAMs was pioneered by Sagiv (44), who employed *n*-octadecyltrichlorosilane to produce both pure and mixed

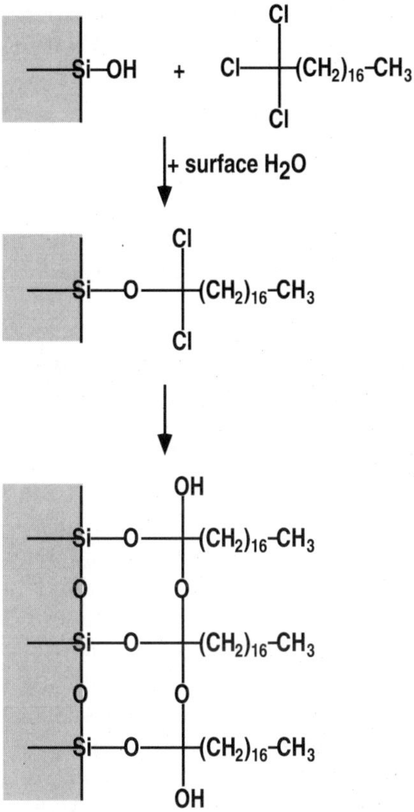

Figure 5.9. Schematic showing formation of an OTS self-assembled monolayer on a silica surface.

highly ordered and packed monolayers on glass surfaces (Fig. 5.9). Alkoxysilanes may also be employed, and both types of reactive silane coupled to glass, quartz, silica, and certain hydrated metal oxide surfaces. In general terms, a small quantity of surface-adsorbed water is required for the reaction to proceed, but care should be taken to exclude excessive water from solution, otherwise self-polymerization can occur, resulting in amorphous multilayer cross-linked networks. Recent work has confirmed that the reaction mechanism is a three-step process, involving hydrolysis of the chloro or alkoxy group to yield a silanol, condensation with surface silanol or metal hydoxy groups, and subsequent cross linking of adjacent silanes through the remaining chloro or alkoxy groups (45). The density of the resulting SAM (and hence the surface wettability) may also be controlled through both the choice of solvent used for the deposition and the use of mixtures of silanes having different chain lengths. Suitable solvents include toluene, hexadecane, and bicyclohexyl.

Another way of achieving SAMs is through the chemisorption of alkanethiols to gold, silver, or copper surfaces. Dialkanethiol and disulphide SAMs may also be formed on gold surfaces. Interest in such films was largely initiated by the work of Nuzzo and Allara (46), although the strong interaction between sulphur compounds and gold had been known since the 1940s. Typically, a gold surface is cleaned until it is hydrophilic and then exposed to a dilute solution ($\approx 10^{-3}$ M) of the organosulphur compound in an organic solvent such as ethanol. Exposure times vary considerably from a few minutes or hours for thiols up to several days for disulphides. Suitable methods for cleaning the gold films include plasma cleaning or a chloroform–methanol wash, followed by treatment with strong aqueous acid. Plasma cleaning alone may not yield consistent monolayers (47). This may be a result of either a metastable gold oxide layer or incorporation into the surface of oxidized hydrocarbon contamination. The cleanliness of the resulting surface may readily be checked using a water drop, which should wet the surface. Whereas Au–S chemisorption has been regarded as involving either an ionic thiolate or covalent but slightly polar bond, fatty acids can form an ionic carboxylate bond on lightly oxidized silver or aluminum surfaces (48,49). As with the organosulphur compouds, this requires a freshly prepared, clean surface which is then exposed to a solution of the fatty acid in a solvent such as hexadecane.

As mentioned above, SAMs may form the chemically selective coating or, using α,ω-bifunctional long-chain molecules, can be used as a way of functionalizing the surface for subsequent attachment of the chemically selective species. Of particular interest in this respect is the work of Wulff and others in which template synthesis was used to produce surfaces capable of reversible host–guest interactions (50,51). In the template approach, the surface modifier is first reacted with a bifunctional molecule that acts as the template to space the modifiers on the surface. This template molecule is then cleaved from the modifiers following immobilization on the surface. Residual silanol groups on the surface are then capped using a reagent such as hexamethyldisilazane. Some examples are shown schematically in Figure 5.10. It should perhaps be mentioned here that the template method was originally developed to produce polymeric materials capable of showing selective host–guest interactions. Such materials have since found application in chromatography, and are capable of separating racemic mixtures. These materials also show clear potential for application to specific chemical sensing.

An alternative approach is to use an ω-functionalized alkyltrichlorosilane. The terminal group can be a protected hydroxyl, ester, or ethylene. Once the

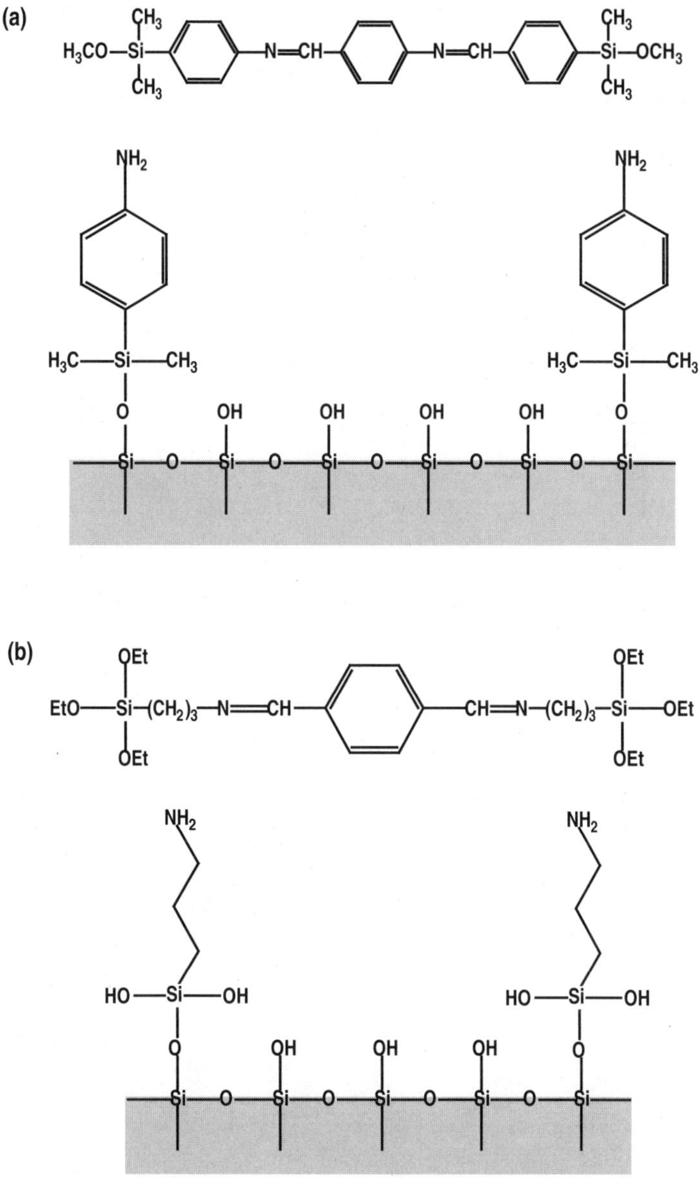

Figure 5.10. Molecular templates and resulting modified silica surfaces. From (*a*) ref. 50 and (*b*) ref. 51.

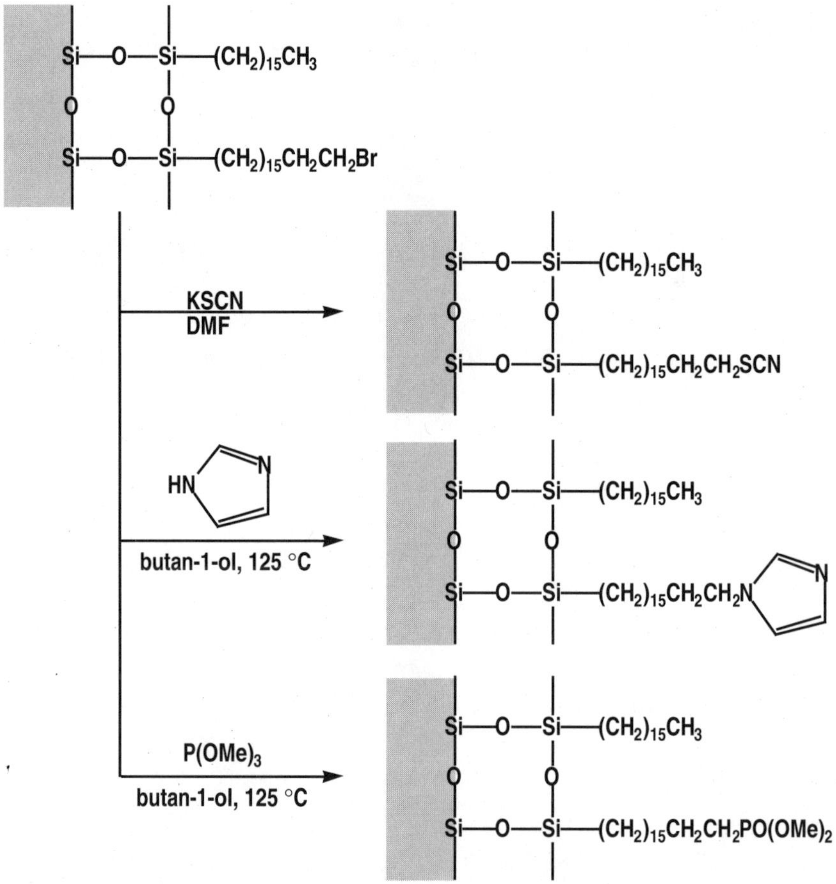

Figure 5.11. Mixed monolayers of α, ω-functionalized silanes and subsequent surface reactions (53).

SAM has been formed, any protecting group is removed and a bilayer formed through covalent binding of a second SAM (52). An alternative approach is that of Rieke et al. (53), in which reactive silanes with a terminal bromo group are incorporated into mixed SAMs. The resulting surface can then be directly modified via a suitable reaction (Fig. 5.11). A similar approach has been employed by Crooks and co-workers using α,ω-carboxy thiols (54,55). This work is of particular interest since a SAW delay line having a gold film deposited between the IDTs was employed to allow real-time monitoring of both film deposition, surface reaction, and subsequent use as a chemical sensor. In one case, 11-mercaptoundecanoic acid was used to form a carboxy-terminated SAM to bind Cu^{2+} ions, the resulting surface being used to selectively detect the nerve-agent simulants diisopropyl methylphosphonate and dimethyl methylphosphonate. In another case, 3-mercaptopropionic acid was used to study hydrogen bonding in bilayer structures formed in the same way.

5.3.4. Covalent Attatchment

There are many other methods of covalently modifying surfaces, including a large number of different chloro- and alkoxysilanes other than those listed above. The main distinction between these and the SAMs produced by OTS, for example, is the resulting organization and packing density. Two of the most widely used materials are aminopropyltriethoxysilane (APTES) and glycidoxypropyl-trimethoxysilane (GOPMS). Such materials have been studied extensively for use as surface modifiers for silicon–silicon oxide, silica and silicate, quartz, glass, metal oxide, and even some polymeric substrates, particularly for the immobilization of proteins such as enzymes and antibodies and the production of chromatographic stationary phases (56–60).

Typical conditions for coupling to a silica or metal oxide surface require the use of a 5–20% solution of freshly distilled silane in dry toluene. In addition, a certain amount of surface-physisorbed water is required, as was the case for the long-chain reactive silanes discussed above. If the solution is not freshly distilled and dry, cross-linking of these di- and trialkoxy silanes occurs in solution prior to coupling to the surface. This can result in loosely bound polymeric particles on the surface that are easily removed. The APTES in particular readily self-polymerizes to form multilayer, cagelike structures on the surface. This is supported by the stoichiometry of the reaction between surface silanol or hydroxy groups and such molecules which has been reported as 1.5–1.6 (61). It is also known that surface water and curing at temperatures above 150°C result in very few unreacted residual silanol or alkoxy groups (60). The resulting polymeric films containing known functional groups can therefore be used either as the SAW sensor coating itself (13,62), or as a means for further immobilizing other species (10,63).

5.3.5. Inorganic Coatings

Although most of the SAW applications described in the literature exploit organic coating materials, a number exploit metals, metal oxides, sol-gel films, and zeolites. Metal and metal oxide films can readily be deposited using thermal evaporation or sputtering techniques, a mask being used to prevent

TABLE 5.4. Examples of Inorganic Film Deposition on SAW Devices

Film	Application	Ref.
Copper	Conductivity	64
Nickel	Conductivity	64
Palladium	H_2	65
Platinum	NH_3	66
$PtCl_2$-amine-olefin	Olefins	72
Silicate sol-gel	Film properties	2, 67
Tungsten trioxide	H_2S	68
Zeolite	Organic vapors	69

shorting out of the IDTs. Examples have included copper, gold, nickel, palladium, platinum, and tungsten trioxide (Table 5.4). Sol-gel films (70) have been used by the group at Sandia National Laboratories (2,67,71). In general terms, a sol is an aqueous colloidal dispersion of solid oxide particles formed by the hydrolysis of a suitable precursor. A typical sol might be formed from tetrathoxy silane in a water–ethanol mixture with trace hydrochloric acid, for example. As polymerization of the silane continues, a continuous solvated cross-linked silicate network is formed—the gel. Solvent is allowed to evaporate under controlled conditions until the required viscosity is obtained, and the film is then deposited by dip coating at a fairly slow rate with a subsequent annealing at elevated temperature (500°C). Zeolite films have also been prepared simply by incorporating the zeolite into a sol-gel coating during processing (69). Other films have been prepared using polymeric materials to entrap the inorganic phase in a thin membrane on the surface of the SAW device. An example is the work of Zellers' group on *trans*-$PtCl_2$(olefin)(amine) complexes for styrene vapor sensing (72).

5.3.6. Physical Entrapment

This technique is commonly employed in biosensors and bioreactors to immobilize species such as antibodies and enzymes. In essence, the molecularly recognitive component is trapped within a cross-linked medium which is permeable to the analyte of interest. We have already touched on some examples, including the use of a zeolite–sol-gel combination and the *trans*-$PtCl_2$(olefin)(amine) complexes mentioned above. Another example is the use of a bisphenol-A matrix to trap triphenylmethane dyes for organic vapor sensing (73). The principal advantage proposed for this method is that selective binding species, especially biomolecules, can be immobilized without the loss of binding activity that is often observed for covalent and physisorption strategies. Another advantage often cited is that the resulting matrix is more robust and stable than covalently bound or physisorbed films. In practice, however, a disadvantage can be the slow response and recovery times. This arises because the pore size has to be small enough to prevent leaching of the binding component, which also reduces the rate of analyte diffusion in and out of the matrix. It is also highly possible (especially with biomolecules) that, over time, the matrix itself will contribute to a loss of activity by the binding component.

Several approaches can be employed. One is to simply mix the required compound with a suitable monomer and to initiate polymerization around it. The possibility remains, however, that the entrapped species will leach from the matrix with time, especially if a large pore size is required for reasons of response time. Another approach is to "label" the binding component with polymerizable functional groups. This can then either be self-polymerized or incorporated in a loose network by copolymerization with a second monomer. Interested readers will find ample examples of these approaches in the literature (74).

5.4. SAMPLE GENERATION AND HANDLING

It should be obvious from the discussion of sensor response that the generation of test and calibration samples, together with their handling and presentation to the SAW device, are as important as the choice of sensor coating. The type and complexity of the systems required for sample generation and handling will depend, to some extent, on the nature of both the sample and the required measurement. In most cases, however, the system will consist of a carrier gas stream, a source of sample gas or vapor, mixing and dilution stages, and provision for both temperature and pressure control. A useful source of information in this respect for gas-phase samples is the book by Nelson (75).

There will be some differences in apparatus between static and dynamic measurements (Fig. 5.12). In the former, for example, measured volumes of gas or liquid samples can be introduced directly into the sensor chamber by means of a valve or injection port. Sample concentrations are then determined by the amount of material introduced to the sample chamber. Static measurements will usually require a means of evacuating the sample chamber and/or purging with nitrogen, helium, or argon gas. For dynamic measurements there is a continuous flow of a suitable carrier gas through the system.

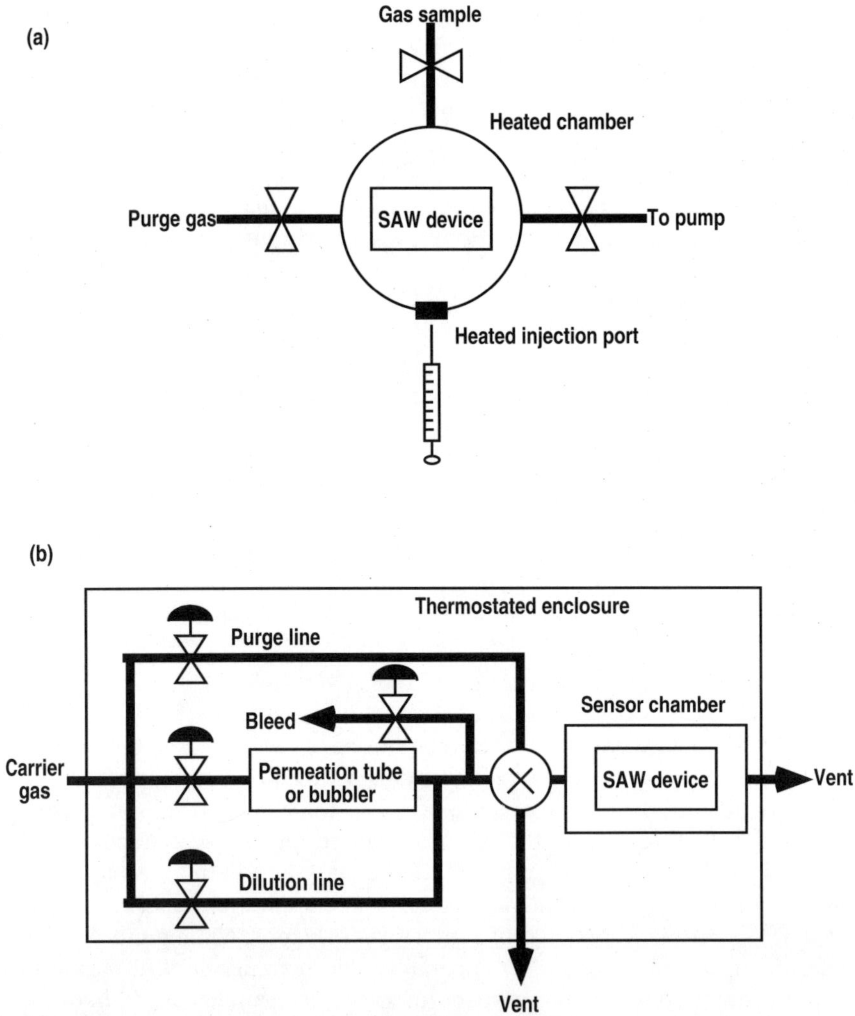

Figure 5.12. Schematic diagrams of gas flow systems for (*a*) static and (*b*) dynamic-sensor response studies.

Sample introduction is then achieved either by stream switching (dynamic steady state) or direct injection (transient). Sample concentrations are then dependent on both the means of sample generation and the use of dilution or mixing stages. Considerations relating to liquid-phase operation are dealt with separately below.

5.4.1. Carrier Streams and Flow Control

As mentioned above, most practical systems require a purge and/or carrier gas stream. In the latter case, gas-flow control and measurement is highly important owing to the kinetic nature of the sensor response. This is especially true for low sample concentrations, because sensor response may show some degree of flow-rate dependence. Flow control is readily achieved using the pressure regulators and check, stream switching, and metering (needle) valves employed in gas-chromatographic systems. Some care needs to be exercised in the materials used in construction of the flow system, especially for reactive gases or organic vapors. In particular, copper tubing and brass connectors should only be used on the upstream side of the system; sample generators and the downstream components should be glass, acid-washed stainless steel, or a suitable inert plastic such as PTFE or PEEK. The use of metal enclosures offers the dual advantages of thermostating and shielding of stray r.f. interferences. It is important, however, to include heat-exchange coils on the carrier gas supply lines within the enclosure to ensure proper thermal equilibration.

Accurate measurement of flow rates is essential for sample dilution for calibration purposes. Traditional volumetric measurements make use of rotameter flow gauges. While simple to install and use, these have the disadvantage of being subject to variations in ambient pressure and temperature conditions. An attractive alternative is the use of mass flow meters, which are often employed in conjunction with a solenoid-operated metering valve for accurate mass flow control. Although details vary, mass flow meters all employ a narrow tube that is heated under constant power. Gas flowing through the tube carries heat energy from one end to the other, resulting in a temperature difference across the tube that is proportional to the specific heat capacity of the gas and its mass flow rate. Mass flow meters are typically calibrated for a specific gas and flow range to provide direct readout in standard units. They can also be used for different gases if the appropriate conversion factors (calculated from thermodynamic considerations) are applied. For low vapor concentrations, these corrections can be neglible compared to the measurement uncertainty, and can therefore be omitted. Experimental flow rates reported for SAW sensor studies vary, but fall in the range from 10 to 500 mL min^{-1}.

5.4.2. Gas and Vapor Samples

Gas samples are in some ways the easiest to handle. In static systems, the concentration of a sample component in an ideal gas mixture can be determined from either the volume or partial pressure ratios. For nonideal gases, a correction factor can be calculated from the ideal and actual gas densities, although at room temperature and atmospheric pressure the correction may be negligible. For ideal mixtures in dynamic systems, concentrations can be determined from the volumetric flow-rate ratio of the component of interest to the total flow through a single mixing–dilution stage, that is,

$$C\ (\text{vol\%}) = \frac{q_s}{q_T} \times 100 \tag{5.9}$$

where q_s is the sample stream and q_T is the total volumetric flow rate. For low concentration, single-component mixtures ($C < 5000$ ppm), this can be simplified to

$$C\ (\text{ppm}) = \frac{q_s}{q_D} \times 10^6 \tag{5.10}$$

where q_D is the diluent flow rate.

Test vapors can also be generated from the corresponding liquid or solid. In static systems, for example, this can be achieved by directly injecting a specific volume of the liquid through a heated injection port. The resulting sample concentration can be expressed by volume or, more conveniently, as mass per unit volume. Assuming ideal behavior, vaporizing a mass m of a substance with molecular weight M_W in a volume V_T gives a concentration of

$$C\ (\text{ppm}) = \frac{mRT}{M_W P} \cdot \frac{1}{V_T} \times 10^6 \tag{5.11}$$

For dynamic systems and samples with an appreciable vapor pressure, concentrations close to the saturated vapor pressure may be obtained by bubbling the carrier gas through the corresponding liquid (Fig. 5.13). For typical organic solvents such as acetone, a single tube may be sufficient. For higher boiling liquids, however, it can be advantageous to connect up to three tubes in series. If low concentrations are required for highly volatile substances, it may be more convenient to blow the carrier gas across the opening of a liquid reservoir.

Another method commonly employed with both solids, liquids, and liquefied gases for low concentrations is the permeation tube (Fig. 5.14). The sample of interest is held in a tube made of a semipermeable material, which may be heated. The carrier gas is blown around the outside of this tube, the vapor concentration being determined by the rate of vapor diffusion through the tube wall.

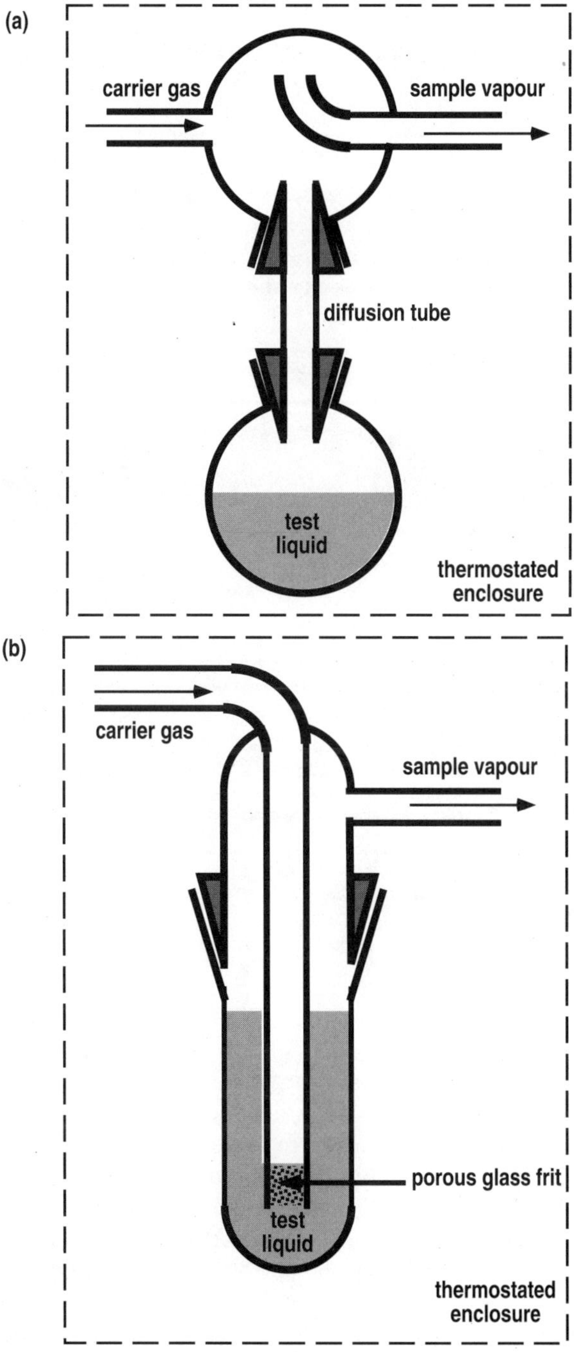

Figure 5.13. Vapor generators for liquid samples. (*a*) Diffusion-tube apparatus for low concentrations and/or highly volatile samples. (*b*) Bubbler employed for typical solvents.

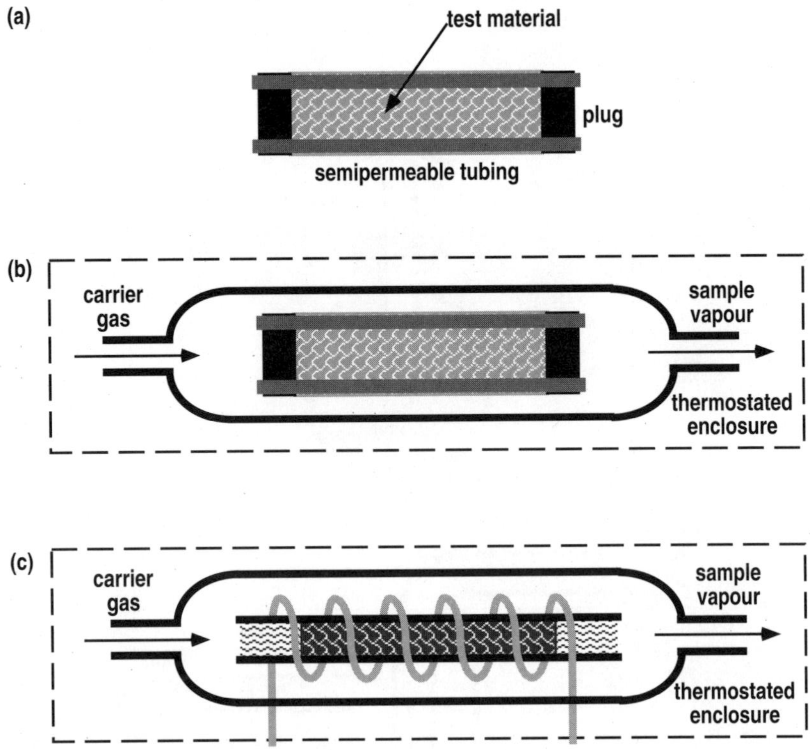

Figure 5.14. Permeation generators for handling gases, liquids, and solids. (*a*) Typical device using semipermeable tubing to house the sample. (*b*) Use of a permeation tube for vapor generation. (*c*) Alternative approach for highly involatile materials using a solid, electrically heated tube with glass wool plugs.

An important issue for dynamic systems is proper calibration of the test sample vapor concentrations. There are a number of possible approaches, depending on the nature and concentration of the particular test compound, but all require stable, accurately measured flow rates. For evaporative methods, the most common method is probably gravimetric: either the sample reservoir is weighed before and after flow of the carrier gas for a fixed period of time, or the sample stream can be passed through one or more cold traps at constant flow rate for a measured time period. With relatively high concentrations and single-component samples, the condensate can then be weighed. For low-concentrations and/or multicomponent samples, the condensate can be dissolved in a known volume of a suitable solvent and the resulting composition determined by gas chromatography. It is also possible to calculate concentrations based on weight loss, vapor pressure and diffusion constants for permeation and diffusion tubes (75). It should be noted, however, that concentrations based on weight loss from a sample reservoir may well be different from actual sensor chamber concentrations if there is significant adsorption or permeation of the sample vapor to or through the walls of the connecting tubing and chamber. Finally, an example of a fully automatic system (including data collection) may be found in the work of Grate et al. (76).

5.4.3. SAW Detector Housings

Some thought must be given to the way in which the SAW device(s) are mounted and coupled to the sample-handling system. The key considerations are precise control of temperature, satisfactory electrical connections, and chemical stability of the housing materials. For example, both brass and aluminum have the advantage of good thermal properties (for heat sinking), are readily machined, and will provide electrical shielding of the device. They are also, however, both prone to corrosion through chemical attack, which is obviously a serious disadvantage for a chemical sensor system! The preferred materials therefore are those already mentioned, such as stainless steel. PTFE has a tendancy to "creep," or deform, under applied stress (such as that arising from clamping screws), and is difficult to machine for the same reason. This can be overcome without compromising resistance to chemical attack by using glass-filled PTFE or an alternative material such as PEEK.

In terms of temperature control, the method adopted will depend, to some extent, on the physical dimensions of the device used. Various approaches have been adopted, including the deposition of thin-film heating "coils" on the reverse side of the piezoelectric substrate. A particularly ingenious approach was adopted by Ballantine and Wohltjen (77), who incorporated TO-8 encapsulated devices into a Swagelok fitting surrounded by heating tape. Other available techniques include attaching the sensor housing to a resistively heated block of brass or aluminum and the use of thermoelectric systems.

5.4.4. Gas-Chromatographic Systems

We have already alluded to the fact that SAW chemical sensors have been incorporated into gas-chromatographic (GC) configurations (3–5). This has the advantage that the chromatograph can be used to achieve sample generation and presentation under reproducible conditions of temperature and pressure. The SAW devices may be used either uncoated or with a selective coating; however, this raises important questions with respect to sensitivity and limit of detection (LOD).

Surface acoustic wave sensor response theory indicates that vapor sorption will increase with increasing film thickness and decreasing temperature. At the same time, however, these factors will contribute to increased band broadening, with a corresponding decrease in peak height and loss of peak resolution (separation between consecutively eluting chromatographic peaks). A related problem is the quadratic nature of the dependence of the resonant frequency on temperature for ST-quartz (Section 3.2.1). This predicts that increased detector temperature will result in rapidly increasing baseline noise because of the finite variation in oven temperature (typically $\pm 0.1-0.5$°C). Several approaches can be adopted in the face of these constraints: (1) thermostat the SAW detector at close to room temperature; (2) use only thin films or uncoated devices; (3) use the largest possible sample volumes with the highest efficiency columns; (4) identify orientations of piezoelectric substrates that have a zero (or very low) coefficient of frequency with temperature at elevated temperatures; and (5) use high-frequency devices ($>$100 MHz).

All of these options, however, involve some degree of compromise. Higher-frequency devices result in lower active areas, for example, so that the expected gain in analytical (as opposed to mass) sensitivity may not be obtained. Likewise, longer chromatographic columns will result in lower peak maximum vapor-phase concentrations passing over the device, which will also reduce detector sensitivity. Thermostating the SAW detector at low temperature also introduces a strong thermal gradient across the device surface due to the temperature differential with the chromatographic column. Watson and Staples (5) recorded a gradient of over 300°C mm^{-1} with the column at 120°C and a SAW device thermostated to room temperature, for example.

From a practical perspective, the best option would appear to be the use of high-frequency SAW resonators, either coated or uncoated, combined with high-precision temperature control. There are two reasons for this. First, even though the expected gain in analytical sensitivity may not be obtained, it will still represent a significant improvement over the use of lower-frequency devices. For example, Wohltjen and Dessy (3) reported minimum detectable quantities of 10 and 1 μg for uncoated 34 MHz quartz and 67 MHz $LiNbO_3$ delay lines, respectively. Watson and Staples, however, report femtogram resolution using uncoated 925-MHz quartz resonators. Second, the reduced device size greatly reduces the dimensions of the detector housing required, reducing the effect of extra-column band broadening and so improving overall resolution and sensitivity.

5.4.5. Liquid-Phase Measurements

With the increasing use of FPW, SH-APM, and STW acoustic sensors, liquid-phase measurements are becoming more common. Sample-handling techniques here are essentially the same as for liquid chromatography (LC) and flow-injection analysis (FIA). A primary concern with respect to any liquid-phase acoustic-wave sensor system is the issue of pressure and flow transients. The pressure-dependence of SAW response, for example, largely rules out the use of the high-pressure reciprocating pumps employed in HPLC. This is especially true for the thin-membrane FPW devices. The peristaltic pumps commonly used in FIA also pose a problem since, by virtue of their operation, pressure (and therefore flow rate) transients are introduced into the system. This will introduce additional noise to the baseline signal unless steps are taken to provide adequate pulse damping. Viable alternatives, then, appear to be gravity feed and gas pressure or syringe pumps.

As far as sample introduction and connections, considerations are the same as for chromatographic systems. These include the prevention of air bubbles entering or forming within the flow system, minimization of connections, and so on, to reduce sample-zone broadening, and careful choice of materials to avoid sample contamination and carryover. This latter point is particularly important for samples of biological origin such as amino and nucleic acids, proteins, and enzymes. Appropriate materials range from PEEK and acid-washed stainless steel to titanium. Proper degassing of solvents and the use of injection loops and stream-switching valves is therefore recommended. In addition, a wide range of different optical and electrochemical detection systems can be run either in series or parallel with the acoustic wave-detection system.

Finally, a wide variety of techniques developed for both LC and FIA allow different concentration profiles and gradients to be generated reproducibly. In combination with stopped-flow techniques, this allows static, dynamic steady-state, transient steady-state, and gradient studies to be performed, a fact that has yet to be exploited in SAW chemical sensor research.

REFERENCES

1. G. C. Frye, S. J. Martin, and A. J. Ricco, *Sensors and Materials,* **1989,** *1,* 335.
2. A. J. Ricco, G. C. Frye, and S. J. Martin, *Langmuir,* **1989,** *5,* 273.
3. H. Wohltjen and R. Dessy, *Anal. Chem.,* **1979,** *51,* 1465.
4. M. Thompson and D. C. Stone, *Anal. Chem.,* **1990,** *62,* 1895.
5. G. Watson and E. Staples, *IEEE Ultrason. Symp. (Proc.),* **1990,** *1,* 311.

6. M. J. Jaycock and G. D. Parfitt, *Chemistry of Interfaces,* Ellis Horwood, Chichester, UK, 1981, pp.12-13, 163ff.
7. C. Tanford, *The Hydrophobic Effect: Formation of Micelles and Biological Membranes,* Wiley, New York, 1973.
8. S. Brunauer, *The Adsorption of Gases and Vapours,* Oxford University Press and Princeton University Press, Princeton, NJ, 1945.
9. S. Brunauer, L. S. Deming, W. S. Deming, and E. Teller, *J. Am. Chem. Soc.,* **1940,** *62,* 1723.
10. D. Li and B. I. Swanson, *Langmuir,* **1993,** *9,* 3341.
11. D. S. Ballantine, S. L. Rose, J. W. Grate, and H. Wohltjen, *Anal. Chem.,* **1986,** *58,* 3058.
12. J. W. Grate, A. Snow, D. S. Ballantine, H. Wohltjen, M. H. Abraham, R. A. McGill, and P. Sasson, *Anal. Chem.,* **1988,** *60,* 869.
13. W. M. Heckl, F. M. Marassi, K. M. R. Kallury, D. C. Stone, and M. Thompson, *Anal. Chem.,* **1990,** *62,* 32.
14. J. W. Grate and M. H. Abraham, *Sens. Actuators B,* **1991,** *B3,* 85.
15. J. W. Grate, M. Klusty, R. A. McGill, M. H. Abraham, G. Whiting, and J. Andonian-Haftvan, *Anal. Chem.,* **1992,** *64,* 610.
16. S. J. Patrash and E. T. Zellers, *Anal. Chem.,* **1993,** *65,* 2055.
17. J. W. Grate, S. J. Patrash, and M. H. Abraham, *Anal. Chem.,* **1995,** *67,* 2162.
18. E. T. Zellers, S. A. Batterman, M. Han, and S. J. Patrash, *Anal. Chem.,* **1995,** *67,* 1092.
19. J. W. Grate and R. A. McGill, *Anal. Chem.,* **1995,** *67,* 4015.
20. J. W. Grate, S. W. Wenzel, and R. M. White, *Anal. Chem.,* **1991,** *63,* 1552.
21. A. W. Snow, W. R. Barger, M. Klusty, H. Wohltjen, and N. L. Jarvis, *Langmuir,* **1986,** *2,* 513.
22. D. L. Bartley and D. D. Dominguez, *Anal. Chem.,* **1990,** *62,* 1649.
23. J. G. Brace, T. S. Sanfelippo, and S. G. Joshi, *Sens. Actuators,* **1988,** *14,* 47.
24. H. Wohltjen, D. S. Ballantine, and N. L. Jarvis, *Chemical Sensors and Microinstrumentation,* R. W. Murray, R. E. Dessy, W. R. Heineman, J. Janata, and W. R. Seitz, Eds., *ACS Symp. Ser.,* **1989,** *403,* 157.
25. W. R. Barger, M. A. Klusty, A. W. Snow, J. W. Grate, D. S. Ballantine, and H. Wohltjen, in *Sensor Science and Technology,* B. Schumm, C. C. Liu, R. A. Powers, and E. B. Yeager, Eds., *Electrochem. Soc. Proc.,* **1987,** *87-15,* 198.
26. C. G. Fox and J. F. Alder, *Anal. Chim. Acta,* **1991,** *248,* 337.
27. A. W. Snow, L. G. Sprague, R. L. Soulen, and H. Wohltjen, *Polymer Preprints,* **1989,** *30,* 213.
28. N. L. Jarvis, J. Lint, A. W. Snow, and H. Wohltjen, in *Fundamentals and Applications of Chemical Sensors,* D. Schuetzle and R. Hammerle, Eds., *ACS Symp. Ser.,* **1986,** *309,* 309.
29. J. A. Groetsch and R. E. Dessy, *J. Appl. Polym. Sci.,* **1983,** *28,* 161.

30. H. Wohltjen, *Sens. Actuators,* **1984,** *5,* 307.
31. A. Snow and H. Wohltjen, *Anal. Chem.,* **1984,** *56,* 1411.
32. M. S. Nieuwenhuizen and A. J. Nederlof, *Sens. Actuators B,* **1990,** *B2,* 97.
33. T. Nomura, T. Yasuda, and S. Furukawa, *IEEE Ultrason. Symp. (Proc.),* **1993,** *1,* 417.
34. J. W. Grate and M. Klusty, *Anal. Chem.,* **1991,** *63,* 1719.
35. G. Roberts, Ed., *Langmuir-Blodgett Films,* Plenum Press, New York, 1990.
36. M. S. Nieuwenhuizen and A. W. Barendsz, *Sens. Actuators,* **1987,** *11,* 45.
37. M. Rapp, D. Binz, I. Kabbe, M. Von Schickus, S. Hunklinger, H. Fuchs, W. Schrepp, and B. Fleischmann, *Sens. Actuators B,* **1991,** *B4,* 103.
38. S.-M. Chang, B. Ebert, E. Tamiya, and I. Karube, *Biosens. Bioelectron.,* **1991,** *6,* 293.
39. S.-M. Chang, E. Tamiya, and I. Karube, *Biosens. Bioelectron.,* **1991,** *6,* 9.
40. G. G. Roberts, B. Holcroft, A. Barraud, and J. Richard, *Thin Solid Films,* **1988,** *160,* 53.
41. B. Holcroft and G. G. Roberts, *Thin Solid Films,* **1988,** *160,* 445.
42. C.-J. Zhong and M. D. Porter, *Anal. Chem.,* **1995,** *67,* 709A.
43. A. Ulman, *An Introduction to Ultrathin Organic Films from Langmuir–Blodgett to Self-Assembly,* Academic Press, San Diego, CA, 1991.
44. J. Sagiv, *J. Am. Chem. Soc.,* **1980,** *102,* 92.
45. M. E. McGovern, K. M. R. Kallury, and M. Thompson, *Langmuir,* **1994,** *10,* 3607.
46. R. G. Nuzzo and D. L. Allara, *J. Am. Chem. Soc.,* **1983,** *105,* 4481.
47. C. D. Bain, E. B. Troughton, Y.-T. Tao, J. Evall, G. M. Whitesides, and R. G. Nuzzo, *J. Am. Chem. Soc.,* **1989,** *111,* 321.
48. D. L. Allara and R. G. Nuzzo, *Langmuir,* **1985,** *1,* 45.
49. N. E. Schlotter, M. D. Porter, T. B. Bright, and D. L. Allara, *Chem. Phys. Lett.,* **1986,** *132,* 93.
50. G. Wulff, B. Heide, and G. Helfmeier, *J. Am. Chem. Soc.,* **1986,** *108,* 1089.
51. Y.-T. Tao and Y.-H. Ho, *J. Chem. Soc. Chem. Commun.,* **1988,** 417.
52. L. Netzer and J. Sagiv, *J. Am. Chem. Soc.,* **1983,** *105,* 674.
53. P. C. Rieke, B. J. Tarasevich, S. B. Bentjen, G. E. Fryxell, and A. A. Campbell, in *Supramolecular Architecture: Synthetic Control in Thin Films and Solids,* T.Bein, ed., Proceedings of the 201st National Meeting of the American Chemical Society, Atlanta, 1991, ACS Symposium Series, Vol. 499, American Chemical Society, Washington, DC, 1992, p. 61.
54. L. Sun, L. J. Kepley, and R. M. Crooks, *Langmuir,* **1992,** *8,* 2101.
55. L. J. Kepley, R. M. Crooks, and A. J. Ricco, *Anal. Chem.,* **1992,** *64,* 3191.
56. R. K. Iler, *The Chemistry of Silica,* Wiley, New York 1979.
57. K. K. Unger, *Porous Silica: Its Properties and Use as Support in Column Liquid Chromatography,* Elsevier, Amsterdam, 1979.

58. *Chemically Modified Surfaces,* J. J. Pesek and I. E. Leigh, Eds., Proceedings of the Fifth Symposium on Chemically Modified Surfaces, Blue Bell, PA, June 16–18, 1993, Royal Society of Chemistry, Cambridge, UK, 1994.
59. K. M. R. Kallury, U. J. Krull, and M. Thompson, *Anal. Chem.,* **1988,** *60,* 169.
60. G. S. Caravajal, D. E. Leyden, G. R. Quinting, and G. E. Maciel, *Anal. Chem.,* **1988,** *60,* 1776.
61. F. Buzek and J. Rathousky, *J. Colloid Interface Sci.,* **1981,** *79,* 47.
62. F. Benmakroha and J. F. Alder, *Anal. Chim. Acta,* **1995,** *302,* 255.
63. J. E. Roederer and G. J. Bastiaans, *Anal. Chem.,* **1983,** *55,* 2333.
64. A. J. Ricco and S. J. Martin, *Thin Solid Films,* **1991,** *206,* 94.
65. A. D'Amico, A. Palma, and E. Verona, *Appl. Phys. Lett.,* **1982,** *41,* 300.
66. A. D'Amico, A. Petri, P. Verardi, and E. Verona, *IEEE Ultrason. Symp. (Proc.),* **1987,** 633.
67. S. L. Hietala, D. M. Smith, V. M. Hietala, G. C. Frye, and S. J. Martin, *Langmuir,* **1993,** *9,* 249.
68. J. F. Vetelino, R. K. Lade, and R. S. Falconer, *IEEE Trans. Ultrason. Ferroelec. Freq. Control,* **1987,** *UFFC-34,* 156.
69. T. Bein, K. Brown, G. C. Frye, and C. J. Brinker, *J. Am. Chem. Soc.,* **1989,** *111,* 7640.
70. C. J. Brinker and G. W. Scherer, *Sol-gel Science: The Physics and Chemistry of Sol-gel Processing,* Academic Press, Boston MA, 1990.
71. G. C. Frye, A. J. Ricco, S. J. Martin, and C. J. Brinker, *Mat. Res. Soc. Symp. Proc.,* **1986,** *121,* 349.
72. E. T. Zellers, R. M. White, and S. M. Rappaport, *Anal. Chem.,* **1990,** *62,* 1222.
73. F. L. Dickert, G. Bertlein, G. Mages, and W.-E. Bulst, *Adv. Mater.,* **1990,** *2,* 420.
74. K. Mosbach, Ed., *Methods in Enzymology,* Vol. XLIV, Academic Press, New York, 1976, pp. 169ff, 191ff, 195ff.
75. G. O. Nelson, *Gas Mixtures: Preparation and Control,* Lewis Publishers, Chelsea, MI, 1992.
76. J. W. Grate, D. S. Ballantine, and H. Wohltjen, *Sens. Actuators,* **1987,** *11,* 173.
77. D. S. Ballantine and H. Wohltjen, *IEEE Ultrason. Symp. (Proc.),* **1988,** 559.

CHAPTER

6

APPLICATIONS OF ACOUSTIC WAVE SENSORS IN ANALYTICAL CHEMISTRY

It should be apparent from the preceding chapters that surface-launched acoustic wave devices offer a wide variety of possible applications in analytical chemistry. Although early work was concerned almost exclusively with chemical sensor development—that is, binding of an analyte to a selective film deposited on a device—other areas of interest have appeared in recent years. An example is the recognition that these devices respond not only to simple mass changes, but also to other interfacial parameters such as acoustoelectric effects, electrical conductivity, and surface free energy. It was also understood at a relatively early stage that sorption of analyte molecules into polymer films imposed on a device surface could lead to changes in certain film properties such as viscoelasticity. This observation has been turned to advantage in recent years through a number of studies on thin-film chemistry, that do not necessarily involve additional analyte molecules.

With respect to biosensor technology, recent studies have attempted to employ surface-launched wave structures for the sensing of liquid-phase biological entities. This effort has been stimulated by analogous early work on TSM devices, the questionable pioneering work on the SAW device, and the drive to employ more sensitive structures.

Finally, the realization that specific chemical sensors for particular analytes are unlikely to appear, at least in the short term, coupled with the desire to characterize multicomponent samples, (e.g., aromas or odors), has led to the design and development of sensor arrays. In this strategy, the aim is to design a set of like sensors having different coatings to study the nature of particular samples (if not *each* individual component) by various chemometric approaches. Such a configuration has attracted the somewhat dubious label of the "electronic nose" (1).

In this chapter, rather than try to cover all possible studies in this area, we will concentrate on a selected number of examples. A more exhaustive picture of published applications (up to 1995 and part of 1996) can be found by consulting the section on coating materials and methods in Chapter 5, especially Tables 5.2 through 5.4. The final section in this chapter consists of some "crystal-ball gazing" on our part with a view to evaluating what lies in the future for acoustic-wave devices in chemistry.

6.1. GAS-PHASE CHEMICAL SENSOR TECHNOLOGY

Here, we introduce those particular applications that have concerned the direct sensing of gas-phase concentrations of a range of analyte species. In particular, an emphasis is placed on reviewing these studies with respect to the sensitivity and selectivity of the sensor, as well as the dynamics of its response. Obviously, for real-time operation in the manner of an ion-selective electrode, true reversible response is a desirable feature for any future device.

6.1.1. Detection of Inorganic Gases

Although attempts have been made to design and implement acoustic wave sensors for NO_x, SO_2, H_2, H_2S, CO_2, and so on, the field appears to be dominated by descriptions of NO_2 detectors. Additionally, nearly all the efforts have concerned SAW delay line or resonator structures and, for the most part, appeared in the 1980s. As mentioned in Chapter 3, a number of articles from different groups appeared in this time period regarding the detection of NO_x species and, in particular, their potential applications in the analysis of environmental samples and automobile exhaust gases. Not surprisingly, any chemically selective film employed in conjunction with a sensor in these areas must be stable at high temperature and be capable of fast, selective, and reversible response. Accordingly, metallophthalocyanines and the parent proton-containing molecule (Fig. 6.1) have attracted considerable interest for the selective binding of NO_2 in view of their stability and widespread use in the fabrication of chemiresistor devices (2–6). The selective layer can be formed on the device by sublimation or by the Langmuir–Blodgett process (6–8) as discussed in Section 5.3. There is evidence that the interaction of NO_2 with phthalocyanine and derivative species can occur by both charge-transfer and physisorptive processes. Accordingly, if a layer of this selective binding agent is imposed on an acoustic-wave device, the mechanism of attenuation of the frequency response may include both mass and conductivity effects. Indeed an early report on the use of the lead derivative on a $LiNbO_3$ device concluded that conductivity effects predominate (2). Others, however, have concluded from concentration-saturation level data that mass effects are more important (6).

The parameters that are crucial from an analytical chemical viewpoint with respect to NO_2 detection have been investigated in detail by Nieuwenhuizen and co-workers (6–9). The sensitivity of the SAW sensor with various

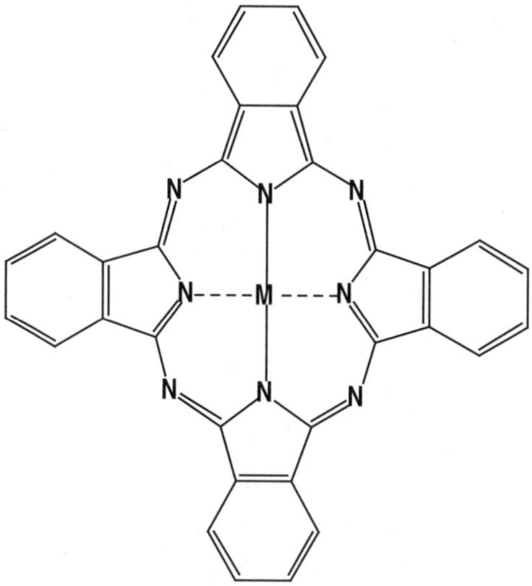

Figure 6.1. Chemical structure of a metallophthalocyanine, where M is the metal atom.

metallophthalocyanines in place (from sublimation), normalized for film thickness, is dependent on device frequency, temperature of operation, and the complexing metal ion. Not surprisingly, decreasing film thickness speeds up response time, but at the cost of lowering the sensitivity. Some of the appropriate data is reproduced in Table 6.1. In terms of the nature of the responses, most of the devices generate signals that are reversible, but considerable variability was found in response times. For 80% "on" and "off" signals, the range was from 2 to 47 minutes for on times and from 11 to 32 minutes for off times. Finally, with regard to the chemical selectivity of the sensors for NO_2 in comparison with the other gases such as NH_3, CH_4, SO_2, and H_2O, the authors found negligible interferent response at the chosen operation temperatures of 70 and 150°C. As a final comment one might mention that, despite the promise indicated by the relatively large volume of research performed on NO_2 sensing, there does not yet appear to be widespread application of this technology in real situations.

Given the advances in hydrogen detection that occurred some time ago as a result of the pioneering work of Lundström and co-workers (Chapter 1), it is not entirely surprising that palladium has been employed as a selective film for SAW sensors (10,11). Analytical data recorded for a 300-nm Pd film in place shows that H_2 concentrations in the range $50-10^4$ ppm can be detected, and that the overall response time is on the order of minutes. Interestingly, in this very early work, the changes in frequency of the SAW device were

TABLE 6.1. Sensitivities of Phthalocyanine- and Metallophthalocyanine-Coated SAW Sensors to NO_2 (6)

Metal	Theoretical Frequency (MHz)	Temperature (°C)	Sensitivity[a] (Hz ppm^{-1} μm^{-1})
H_2	78.95	70	49
H_2	78.95	150	75
Mg	78.95	70	44
Mg	78.95	150	45
Fe	78.95	70	452
Fe	78.95	150	67
Co	78.95	70	138
Co	78.95	150	133
Ni	78.95	150	33
Pb	78.95	70	104
Cu	39.48	150	30
Cu	52.63	150	46
Cu	78.95	150	80
Cu	98.69	150	109

[a] Measured on the response-concentration curve at 100 ppm and corrected for film thickness.

interpreted in terms of alteration of the elasticity and density of the Pd film. Although the inventors of this technology were optimistic that their sensor compared favorably with other systems, it does not appear to have been accepted as a routine detector for H_2 at this time.

As was the case for much of the work described above, efforts on SO_2 and H_2S sensing began in the 1980s. Vetelino and co-workers (12) were able to develop a selective $LiNbO_3$-based SAW detector for SO_2 using a film of triethanolamine. Apparently, the sensor operated with a sensitivity of 100 Hz ppm^{-1} and possessed a limit of detection of about 10 ppm SO_2. Moreover, the authors claim to be able to detect concentrations as low as 1 ppm.

Organotransition metal complexes have also been used to attempt the selective binding of SO_2 at a device–gas interface (13). The complex, bis(sulfur dioxide)tetrakis(triphenylphosphine oxide) manganese(II)dioxide, was converted to the mono-SO_2 adduct and coated onto a 52-MHz ST-quartz SAW delay line. On exposure to SO_2, it was noted that the response could be described as quasi-reversible, meaning that the complex containing two SO_2 molecules required reconversion to the monoadduct before further use. In addition, the sensor responded to gas-phase concentrations of 20 mg L^{-1}. A second complex, bis(tribenzylphosphine)copper(II)thiophenolate, also forms an adduct with SO_2, in this case with a single molecule of the S-containing moiety. With this complex, the signal was reversible and the response time for concentrations of SO_2 of 1 g L^{-1} was about 2 minutes. The signal did not return completely to the baseline, however, suggesting that some SO_2 was retained in the complex at the measurement temperature (25°C).

The Vetelino group has presented research on the detection of H_2S (14) using WO_3 as the selective film. The measurement range for the device was 10 ppb to 30 ppm at a temperature of 130°C. The response time was on the order of minutes and, apparently, the detector did not respond to either N_2 or CO.

In summary, with respect to the detection of small molecules, there have been a few forays into sensing of NH_3 and CO in addition to those gases mentioned above. There does not, however, appear to have been a great deal of progress in terms of either additional work or actual applications. Possible reasons for this lie in the difficulties of generating "practical" sensor packages and producing truly selective and stable films for the trapping of molecules. Additionally, these types of detector often have to be placed in deleterious media such as stack gases and exhaust fumes and, finally, must compete with a number of established technologies.

6.1.2. Selective Detection of Organic Vapors

As we have seen, there is a plethora of work on the exposure of devices to a wide variety of organic molecules, particularly those involving polymer films. In this section, we restrict ourselves to examples of research concerned with the detection of specific molecules or groups of organic compounds that have similar structure. This area has been very much the province of Zellers and his group at the University of Michigan although, of course, there have been a number of other significant contributions.

One of the first attempts at the highly selective detection of a particular molecule was that of Snow and Wohltjen (15), who used a coating of poly(ethylene maleate) for the determination of cyclopentadiene. This exploited the Diels–Alder reaction to produce a cyclopentadiene dosimeter. The film was spin-coated onto a 31-MHz ST-X quartz SAW delay line, and the system tested using static measurements in a 2-L flask. Injection of cyclopentadiene into the flask resulted in a steady decrease in SAW resonance frequency due to formation of the Diels–Alder adduct, with the rate of frequency decrease being linearly proportional to concentration over the range 100–10,000 ppm. Flushing of the flask restored a stable baseline, with an irreversible shift corresponding to the increased mass of the film. The interference effects of various organic vapors was also studied. It was concluded that the sensor could detect 200 ppm of cyclopentadiene using an integration time of 1 minute, and that the dosimeter would have a lifetime of ≈1700 hours if average exposures were of the order of 10 ppm.

Bein et al. (16) used zeolitic materials in conjunction with a 97-MHz SAW sensor. Crystals of ZSM-5 zeolite (pore size $0.54 \times 0.56\ nm^2$) were embedded in a glassy silica matrix which was then coated onto the SAW device. The silica is prepared by hydrolysis and polymerization of tetraethylorthosilicate. The SAW sensor with silica–zeolite composite in place gave signals to methane and isopropyl alcohol at levels of 500–800 ng cm^{-2}, although no indication was given as to the calculation of these figures. The corresponding value for 2,2,4-trimethylpentane was <5 ng cm^{-2}. No indication of response time, reversibility, or temperature stability was given.

On a similar theme, the exploitation of host–guest chemistry was also used to good advantage by Dickert and co-workers (17) in their study of tetraazaparacyclophanes (Fig. 6.2). In this work, two aspects were examined: first, molecular modelling (MM3) methods were employed to consider the energetics of host–guest interactions; second, SAW responses with layers of cyclophane in place were measured. The MM3 calculations generated the reorganization energies for intracavitative adducts of cyclophane with aromatic species such as toluene. Essentially, the calculations were capable of a modest prediction of the "fit" of the cavitand into the host molecule. These predictions were then correlated with the responses of a cyclophane-coated SAW device to acetone, dichloroethene, carbon tetrachloride, chloroform, toluene, and tetrachloroethylene. The sensor response was reversible and "fast" in that rise and fall times were on the order of minutes.

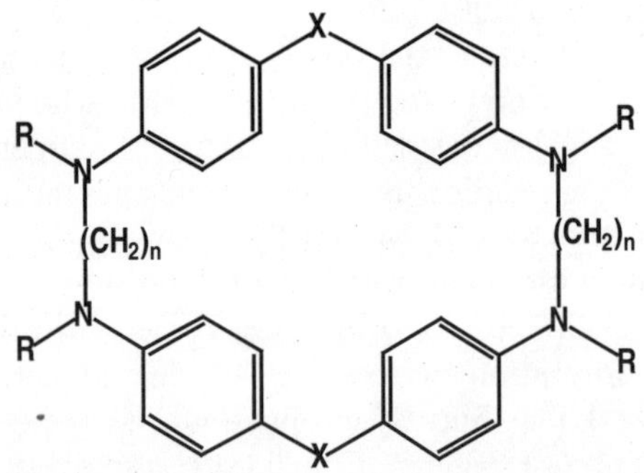

Figure 6.2. Substituted tetraazaparacyclophane [$X = O$, $n = 4$, $R = —(CH_2)_4CH_3$] (17).

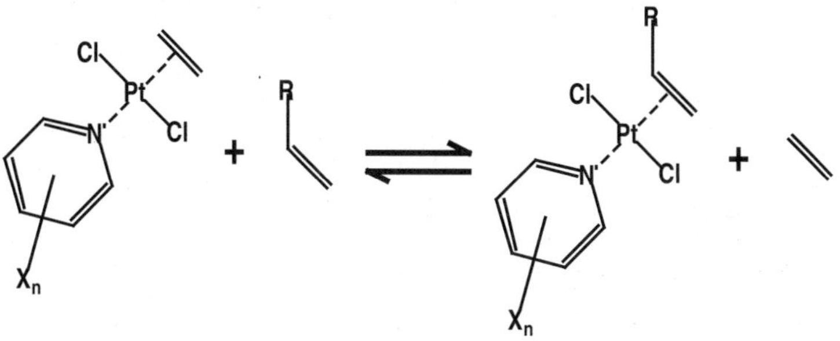

Figure 6.3. Schematic depicting interaction of olefinic species with an (olefin)(amine) platinum complex. X_n indicates variable substitution on the pyridine ring and R = phenyl for styrene vapor (18).

The coating of organometallic platinum complexes on dual-SAW delay-line devices for interaction with various olefins has been emphasized by Zellers and co-workers (18–22). Detection of these molecules on a real-time basis is important from environmental and occupational health standpoints in the fiber-resin-reinforced plastics industry. For example, workplace exposure to styrene vapor is said to lead to a number of health issues, including irritation of various tissues and genotoxicity. The basic scheme for incorporation of selective binding of double bond-containing moieties to the surface of the sensor involves *trans*-$PtCl_2$(ethylene)(pyridine) or *trans*-$PtCl_2$(ethylene)(substituted pyridine) complexes, as shown in Fig. 6.3. With regard to sensor development, the nature of this interaction has been examined in thorough detail.

When a 30-MHz device is used together with a coating of the complex incorporating unsubstituted pyridine and iso-butylene polymer, fast responses and recovery (seconds) on exposure to 5- to 700-ppm concentrations of styrene were observed (19). The effect of increasing operating temperature is to cause a decrease in frequency shift, the slope of the isotherm at each temperature being proportional to the air–coating partition coefficient. At higher styrene concentrations, departure from this behavior is associated with multilayer buildup of the olefin on the sensor surface. The limit of detection of the configuration is in the range of a few parts per million. With respect to the selectivity of the sensor over other molecules, the authors were able to establish that (with the exception of butadiene) a high degree of selectivity to styrene can be achieved (18). Finally, clever manipulation of the structure of the organoplatinum complex can result in the tuning of device selectivity to other olefin species based on steric and electronic factors (20–22).

As a postscript to this section, it is interesting to note that there has been commercial interest in the development of SAW sensors for the detection of substances of abuse (23) and explosives (24). In a novel setup, a SAW resonator device is exposed to the nozzle outlet of a quartz capillary column in order to develop a combined GLC–sensor system (Fig. 6.4). No attempt to increase the selectivity of the sensor component by using binding agents was made in this work. The necessary selectivity can, apparently, be achieved using the chromatographic equipment and temperature control. Employing a 925-MHz resonator, femtogram quantities of compounds such as heroin, cocaine, and tetrahydrocannabinol can be detected on the surface of the sensor. Interestingly, the selectivity of chromatography together with the extremely high sensitivity of the resonator has resulted in attempts to detect extremely low concentrations of various nitro-containing organic compounds (explosives) in air (24). The emphasis here is clearly connected with airport security considerations.

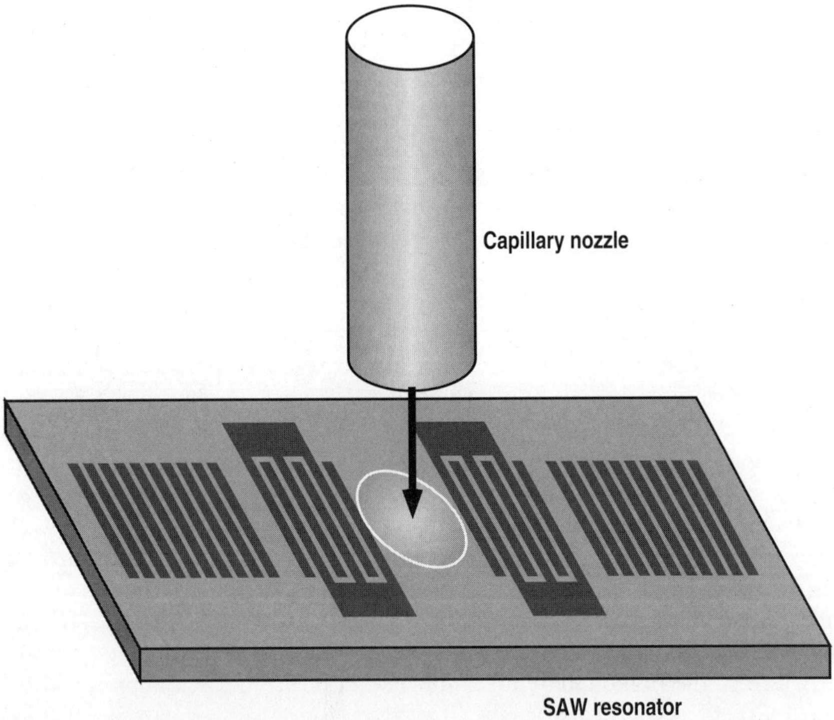

Figure 6.4. Schematic showing the capillary column effluent impinging directly onto the surface of a SAW resonator (24).

6.2. STUDY OF THIN-FILM PROPERTIES

In Chapter 3, we discussed in detail aspects of the interaction of acoustic waves with thin films on device surfaces. This included the physics of film properties from an intrinsic standpoint and from the view of perturbations caused by vapor sorption. To summarize, viscoelastic effects in materials are connected to the complex shear modulus when the matrix is subjected to shear forces (as is the case for acoustic wave devices for materials deliberately placed on the device surface). The modulus is comprised of elastic and loss moduli and is highly frequency and temperature dependent. Accordingly, monitoring these parameters for coated devices can lead to extremely useful information concerning materials, such as glass transition processes in polymers. In this section, we present a few examples of the use of SAW devices for studying thin-film properties.

6.2.1. Polymer Thermomechanical Properties

As a component of the pioneering triad of papers published by Wohltjen and Dessy (25–27), the possibility of using surface acoustic waves to probe the thermomechanical properties of polymer films was introduced. Thin films of polycarbonate, polysulfone, and copolymers of these two materials, as well as polytetrafluoroethylene, poly(methyl methacrylate), and photoresist were placed on the surface of a SAW device for study at different temperatures (0–200°C). Discontinuities in the measured wave amplitude with temperature were interpreted, in general terms, as being associated with the glass transition temperature. Simply put, increased "flow" in the polymer film led to better adhesion to the substrate with the ensuing consequence of significantly increased association of the Rayleigh wave. The authors of this work emphasized that acoustic-wave devices offer the sensitive measurement of glass and other transitions in rather small samples.

6.2.2. Diffusion in Gels

Although the earlier research on using acoustic waves to characterize gels was performed using TSM sensors (28), this strategy was extended to FPW devices by White and co-workers (29). These workers formed 250-μm thick gels from agar, gelatin, poly(vinyl alcohol), and poly(acrylamide) on microfabricated flexural plate-wave delay-line structures consisting of silicon nitride, zinc oxide, and aluminum composite membranes. To explain the interaction between plate waves and the gel films two theoretical approaches, based on impedance and transmission-line methods, were employed. At this time, little or no connection between the experimental behavior of the device with gel layers in place and the predictions made by the aforementioned theories were made. The authors pointed out, however, that such approaches might be useful for examining the process of diffusion in gels, and indeed subsequently went on to perform explicit experiments (30).

Here agarose, agar, and calcium alginate gels were formed on an FPW device which was, in turn, exposed to flowing liquids as shown in Figure 6.5. The resulting frequency–time profiles, $f(t)$, are then a function of the density change resulting from diffusion into the gel:

$$f(t) = \frac{1}{\lambda}\sqrt{\left(\frac{(2\pi/\lambda)^2 B}{M + \rho_f(t)\delta}\right)} \tag{6.1}$$

where λ is the acoustic wavelength, B is the effective plate stiffness, M is the mass per unit area of the sensor plate, $\rho_f(t)$ is the time-dependent gel density, and δ is equal to the term (K/k) in Equation 3.4 (Section 3.1.2). Because acoustic propagation into the gel occurs by an evanescent wave mechanism, δ is the decay length for this process into the gel. Equation 6.1 (which may be derived from Equation 3.4) therefore allows computation of gel-diffusion

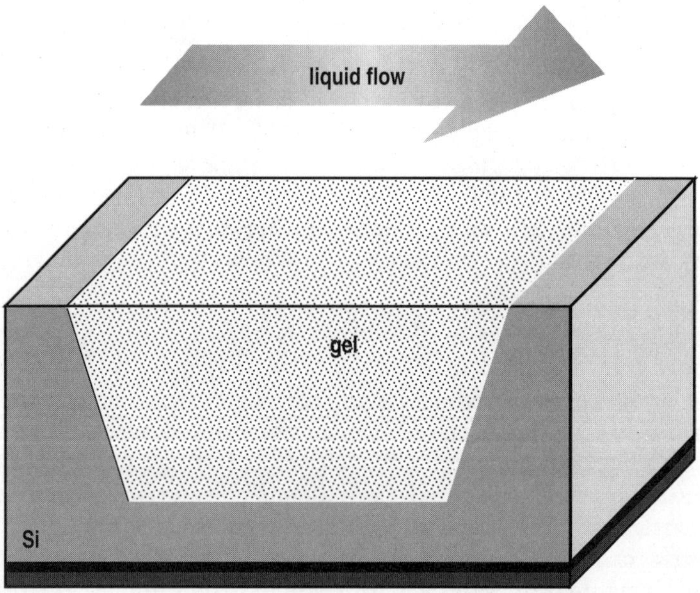

Figure 6.5. An FPW sensor coated with gel and subjected to liquid flow (29, 30).

coefficients from the sensor-response profiles employing a Fick's second law approach and the measured density of the virgin gels before and after their exposure to the sample stream. Interestingly, the agreement between the diffusion coefficients measured by the FPW device and values calculated from a theoretical model was excellent.

6.2.3. Film Process Monitoring

The SH-APM device has been used by Martin and co-workers (31,32) in interesting studies of thin-film photopolymerization and of icing on surfaces. In the former work (31), negative photoresist films were applied to the electroded surface of an ST-quartz SH-APM device. The input and output IDTs consisted of 75 finger pairs having a periodicity of 50 μm. Additionally, an aluminum film 10 nm thick was formed on the wave path of the device, which was operated around 100 MHz. Photoresist films were spin-coated onto the Al film and, after some heat treatment, were subjected to ultraviolet radiation. The device was monitored both with respect to changes in acoustic-wave velocity and amplitude. The former parameter is related to the frequency of operation, whereas the latter is associated with loss of acoustic energy (attenuation) between input and output ports. A theoretical analysis of the interaction of acoustic waves with the photoresist polymer film led to the following expressions for changes in wave velocity (Δv) and attenuation ($\Delta \alpha$):

$$\frac{\Delta v}{v_0} \approx c_m \, \Delta\left(\frac{hG'}{v_0^2} - h\rho_f\right) \tag{6.2}$$

$$\Delta \alpha = \frac{c_m \omega}{v_0^3} \, \Delta(hG'') \tag{6.3}$$

which may be directly derived from Equations 3.24 and 3.25 in Section 3.7.3. If we assume that the thickness and density of the film do not alter during the cross-linking phenomena associated with photoinduced effects, viscoelastic parameters can be gleaned from the APM measurements directly.

The authors worked with two aspects of the photopolymerization experiment, the first being study of the kinetics of velocity and attenuation processes and the second being an examination of the wavelength dependance of cross-linking effects. The effect of cross-linking is to cause an increase in elastic storage modulus (G') and loss of surface mass [N_2 molecules from the photoinitiator, 2,6-bis-(p-azidobenzylidene)-4-methylcyclohexanone]. These processes should both result in an increase in device velocity, which was indeed observed. In contrast, cross-linking is expected to decrease the loss modulus, resulting in a decrease in attenuation, which again was observed. Finally, the device was successful in measuring the spectral sensitivity of the photoresist film.

In the second study (32), a somewhat analogous experiment to that described immediately above was performed on the liquid-to-solid phase transition. Freezing of a liquid is expected to result in an increase in propagation loss due to the change from viscous damping effects to elastic processes in the new phase. Furthermore, the magnitude of the loss will be governed by the level of contact of the film with the device surface. Using similar devices and measuring equipment to those mentioned above, the propagation loss for water, water containing antifreeze, and gallium films were all measured as a function of temperature. A loss of $\approx$6 db was observed for liquid water contacting the device surface, whereas the value for ice was $\approx$12 db, confirming the dramatic increase in loss associated with solid-phase formation. The analogous experiment with elemental Ga showed an initial loss of $\approx$5 dB for the liquid, rising to $\approx$18 dB on solidification. After a short period, however, the film "delaminated" from the device so that the Ga was found to be completely free of the device surface. Taken together these results convinced the authors that the SH-APM device could be employed to monitor the icing of aircraft wings and the behavior of agents employed to prevent crop damage from frost. In neither case was it explained how the sensor could be utilized in practical situations.

6.2.4. Static and Dynamic Glass Transitions of Polymers

Solid-phase transition behavior has also been examined using an FPW sensor. In this case, Grate and co-workers (33) studied changes in polymer properties through the correlation of frequency with temperature. Polymer films were formed by spray coating samples of poly(vinyl acetate), poly(isobutylene), poly(vinyl propionate), and poly(*t*-butyl acrylate) onto FPW devices operated at a frequency of 5 MHz. Essentially, these layers were investigated with respect to two transition effects, that is, the static and dynamic glass transition temperatures, T_g and T_α, respectively. As explained earlier (Section 3.7.2), increasing a polymer's temperature past T_g is associated with a change from a glass- to a rubber-like material. Associated with this process is an increase in free volume. The second parameter (T_α) involves the motion of polymer chains (relaxation processes) and is, thus, often referred to as the dynamic transition. At T_g, a significant change (break downward) occurs in the slope of the frequency-temperature plots for poly(vinyl acetate) and poly(*t*-butyl acrylate). This result is related to the expected abrupt change in acoustic velocity at T_g. Apparently, with respect to T_α, a minimum in wave amplitude and a sigmoidal decrease in the frequency–temperature curve are observed. These features can be related to the dynamic glass transition temperature by a comparison of the polymer relaxation times with the periods of the acoustic waves propagated by the device. When the relaxation time and the wave period are similar, acoustic energy can be more efficiently coupled into the material, causing the observed amplitude minimum. In dealing with the results of these experiments, the authors were at pains to emphasize the consequences of their work for those interested in employing polymer films for "trapping" vapors (see previous section). Important considerations are diffusion processes in glass and rubber materials, the "mixing" of gravimetric and elastic effects, and the actual effect of the adsorbed molecular species on polymer properties.

6.2.5. Sorption Studies with Structured Films

Rayleigh wave sensors have been utilized for the study of thin films with unusual structures such as those involving pore formation (34) and "hairy" buckyballs (Buckminsterfullerene, C60) (35). In the former work, silica particles of ≈20 nm were formed into a mesoporous film (pore size 2–10 nm) on an ST-quartz SAW device by the Stöber method. A microporous film was also prepared by the sol-gel processing of a tetraethoxysilane solution. Classical BET adsorption isotherms for N_2 and CO_2 were compared for the uncoated and coated devices. The results led to the conclusion that sensitive and convenient measurements of adsorption isotherms for very thin films using different gases can be obtained using SAW devices.

In the buckyball study, the derivatized C60 molecule was anchored to the surface of a 200-MHz SAW resonator (Fig. 6.6). It is important to note that the use of the well-known silanization strategy involved led to the formation of a multilayer structure incorporating the C60 moieties. This was characterized by optical absorption and by polarized variable-angle attenuated total-reflection IR spectroscopy and was thought to be about 20–30 nm thick. The coated devices were then exposed to various molecules known to interact with buckyballs, namely, decahydrophthalene, perchloroethylene, and toluene, and to those that do not, such as benzene and carbon tetrachloride. The relatively large changes in SAW frequency associated with the former three molecules were interpreted solely in terms of the known solvating ability of these molecules for C60. Interaction of any of the adsorbed vapor species with the cross-linked siloxane structure and instigated changes in the polymer matrix were not addressed.

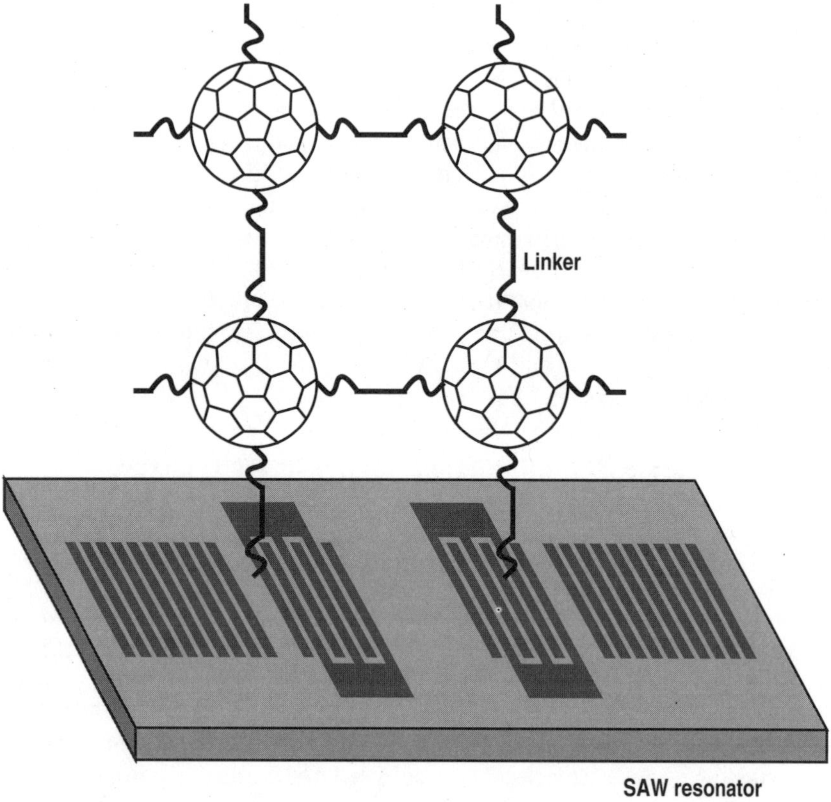

Figure 6.6. Covalently bound buckyballs on a SAW resonator surface (35).

6.3. ACOUSTIC-WAVE DEVICES IN THE BIOSENSOR FORMAT

This field was instigated in an unusual way in that the first report of an SAW sensor to detect real-time immunochemical interactions (36) was later subjected to criticism on technical grounds. Apparently, true Rayleigh wave behavior could not be established for the operation of a SAW in liquids (37) (see Section 3.6). Nevertheless, the first report stimulated interest in the use of acoustic-wave devices to detect, for example, antibody–antigen interactions and nucleic-acid hybridization. Not surprisingly, the area has been heavily dominated by studies using the TSM sensor since the initial work of Thompson et al. (38,39). Work on the TSM sensor for examining immunochemical interactions was highly prevalent at the Fourth World Congress on Biosensors (Biosensor '96) held in Bangkok in that a variety of studies from several groups were reported (40). Analogous investigations with surface-launched acoustic-wave sensors, however, have been relatively few in number. The flurry of activity in the development of SH-APM and FPW devices that occurred in the late 1980s to early 1990s does not yet appear to have been translated into many biosensor applications. Selected examples of those studies that have appeared are described in this section.

An early measurement involving a biochemical system was that of White et al. (41), who were able to detect the adsorption of protein to the surface of an FPW sensor. Additionally, the device was subjected to fluids of various viscosities, including a sample of blood. The delay-line oscillator frequency of the device was monitored subsequent to the introduction of solutions of bovine serum albumin (BSA) to its surface. The frequency drop ($\approx$2 kHz) was attributed entirely to a gravimetric effect resulting in the adsorption of 2.8×10^{-6} g cm^{-2} of protein. Interestingly, this work was one of the first to recognize that acoustic-wave sensors, in this case the FPW device, show great potential for viscometric measurements on biochemical systems. In this respect, the authors pointed out that increasing viscosity of liquids imposed on the FPW device leads to rising attenuation of the plate wave. This effect was used to demonstrate the time dependence of blood coagulation. Although this preliminary study did not involve a biochemically interactive pair, it was important in emphasizing that dual information could be gleaned from FPW device response, that is, gravimetric and viscometric effects.

One of the earlier articles dealing with a genuine biochemical pair was that of Andle et al. (42) concerning the detection of nucleic acid hybridization. Figure 6.7 depicts schematically the attachment of a layer of a single strand of the homopolymeric oligonucleotide of dT (20-mer of deoxythymidylic

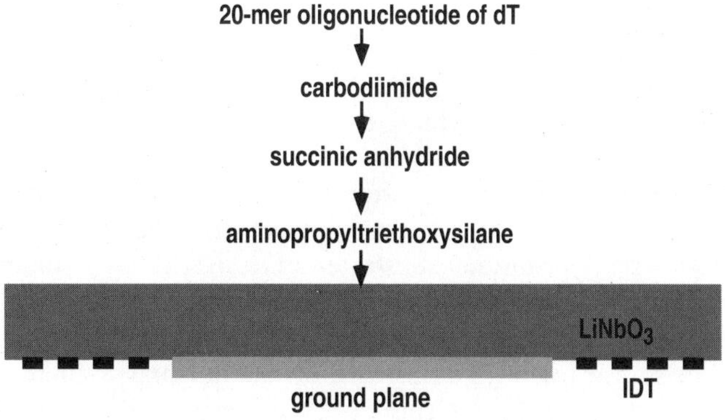

Figure 6.7. Immobilization sequence for the homopolymer dT on a $LiNbO_3$ SH-APM sensor.

acid) to the surface of a $LiNbO_3$ SH-APM sensor operated in the delay-line mode. The device was operated in both single- and dual-delay line configurations, the latter to compensate for changes in temperature, conductivity effects, and so on. The authors chose to measure the relative phase among a number of possible parameters in order to detect chemical changes on the device surface. The device was then exposed to several injections of dA homopolymer (the complementary 900-base sequence of deoxyadenylic acid). In general terms, phase decreases and increases were interpreted in terms of frequency increases and decreases, respectively. Interestingly, the device responses were said to be governed by both viscoelastic (due to hybridization) and "mass" effects. No attempt was made either to characterize the silanization strategy employed for covalent binding of the oligonucleotide to the surface or to detect the amounts of bound material by radiolabelling, a fact which has generally plagued the biosensor field.

In 1992, Lowe and co-workers (43) described an alternative to the conventional SH-APM device described in this text, and demonstrated its use for measurement of protein adsorption to the surface of a thin polymeric layer deposited on the device. The sensor consisted of crystal of Y(42.5°)-cut, Z-propagating quartz as a substrate to support surface-skimming bulk waves at a frequency of 110 MHz. Shear acoustic-waves were propagated in a dielectric overlayer (such waves are called Love waves), and the velocity change of these waves caused by deposited mass determined. The adsorption of IgG to a poly(methyl methacrylate) film spin-coated onto the device surface was then measured though the phase change as described above. Finally, the authors were able to demonstrate binding of anti-IgG antibody to the IgG already adsorbed at the device–liquid interface. It was emphasized that the sensor displays similar sensitivity and response times to other immunosensors, but exactly which structures were being compared was not mentioned. Subsequent to this work, a further article also suggested that the Love wave device could be useful for biochemical sensing, but no biochemically interactive pairs were investigated (44).

In a somewhat unusual experiment, Shiokawa and co-workers (45) were able to use a "leaky" SAW device based on $LiTaO_3$ to detect pH changes associated with biological moieties through the acoustoelectric effect. The enzymes urease and cholinesterase were immobilized on the device surface by classical silanization procedures prior to sensing the interaction of the enzymes with the appropriate substrates. The well-known inhibition of cholinesterase by organophosphorous pesticides was also observeable with this sensor.

The surface transverse wave device introduced in Chapter 2 has also been employed as a biosensor (46). In this case, the sensing area was coated with

50 nm of silica which served two purposes: first, it provided a degree of protection for the aluminum IDTs; second, it provided a greater area of silanol groups for chemical immobilization protocols. In this work, two STW devices were incorporated as two arms of a r.f. interferometer in which one served as the analytical sensor while the other provided a reference signal. As with a number of other devices, the electronic part of the experiment was, in essence, a measurement of the relative phase between the two STW sensors. The procedures employed to measure biosensor signals in terms of the surface chemistry were, however, somewhat unusual in that the antigen (atrazine) for immunochemical detection was attached to the surface of the device rather than the antibody. This was accomplished by chemically labelling the atrazine molecule with biotin. Avidin was then immobilized on the device surface using (γ-glycidoxypropyl)trimethoxy silane and the antigen coupled using the well-known avidin–biotin interaction. Assessment of solution concentrations of atrazine was made by utilizing a competitive assay. Apparently, in this configuration, quite rapid response times (1–3 min) and low limits of detection (ppt) could be achieved. It is interesting to note that the STW sensor has also been employed as a detector for liquid chromatography in terms of separation of biological species (47). Finally, we note that in none of the above studies was the surface chemistry established by quantitative procedures such as the use of surface analytical and/or radiolabelling procedures.

In a recent article, Josse et al. (48) evaluated the use of the SH-APM devices in the development of immunosensors. The device employed in this work was a delay line formed on ZX-$LiNbO_3$ and operated at a fundamental frequency of 51 MHz. Other modes of operation, however, could also be accessed to generate analytical information based on multifrequency response. The experimental configuration adopted consisted of the imposition of surface chemistry on the nonelectroded face of the device and of the use of the usual analytical and reference sensors for correction for temperature and other effects. Electronically, measurements consisted of monitoring the phase shifts and signal amplitudes of the outputs of both lines relative to the input. Two synchronized relays were employed to switch between the analytical and reference channels. Immunoglobulin G (IgG) was immobilized on the surface of the sensor using three strategies involving underlying layers of silane, dextran, and poly(etherurethane) prepolymer (XP-5). The thickness of these films was computed from X-ray photoelectron spectroscopic measurement of the intensity of the substrate Nb 3d electrons. Curiously, the inelastic mean free path of these electrons was found to be 13.6 nm in dextran and 3.2 nm in XP-5. Unlike the studies mentioned above, this investigation employed standard ELISA protocols to calculate quantities of antigenic moieties bound

TABLE 6.2. Antigen Binding Capacity and Sensitivity of SH-APM Device to Immunochemical Reactions (48)

Immobilization Protocol	Binding Capacity ($ng\ nm^{-2}$)	Sensitivity ($Hz\ mm^2\ ng^{-1}$)[a]		
		50.6 MHz	151.1 MHz	152.6 MHz
Aminosilane	0.82	210	800	470
XP-5	2.2	57	290	120
Dextran	3.7	83	170	130

[a] Calculated from ELISA measurements and corresponding frequency-shift data.

to the IgG immobilized by the various layers. The immunosensors were then exposed to immunochemical reactions in a no-flow cell in order that frequency shifts could be measured (from phase differences). Elements of the data are summarized in Table 6.2. One important observation was that the expected dependence of the response (Δf) on the unperturbed resonance frequency (f^2) was not found when the response of the dextran-coated immunosensor was recorded at the fundamental and third harmonic frequencies. From this, the authors raised the important issue of the passage of acoustic waves through the underlying layers used to immobilize the biological entity. In particular, viscoelastic effects are prevalent in hydrogel films, which may also be perturbed by the immunochemistry. Accordingly, this article is one of very few that have delineated the inherent dangers that can occur from the behavior of the "cocktails" used to attach biological species to device surfaces.

6.4. SENSOR ARRAYS AND PATTERN RECOGNITION

As mentioned above, the design and generation of a specific sensor for a particular analyte in a gas-phase mixture of compounds is an extremely difficult task. For a number of analytical applications, such as various foodstuffs, however, it is only necessary to recognize odors or aromas, that is, a complex set of molecules, as an *entity* without necessarily picking out each individual molecule. In this section, we describe work combining sensor arrays with mathematical approaches to pattern recognition that have been employed to address such problems. First, however, it is necessary to develop some important criteria and definitions regarding the nature of chemical sensor response.

6.4.1. Chemistry and Physics of Analytical Selectivity, Specificity, and Sensitivity

In a seminal article, Kaiser (49) set out careful definitions of the parameters delineated in the title of this section. A summary of this treatment is presented here. For a multicomponent sample being analyzed by a particular technique (discussed below), a rigorous approach based on Equation 2.31 can be instigated. For an n-component sample, a system of *calibration functions* can be specified:

$$\begin{aligned} x_1 &= \gamma_{11}c_1 + \gamma_{12}c_2 + \cdots + \gamma_{1n}c_n \\ x_2 &= \gamma_{21}c_1 + \gamma_{22}c_2 + \cdots + \gamma_{2n}c_n \\ x_n &= \gamma_{n1}c_1 + \gamma_{n2}c_2 + \cdots + \gamma_{nn}c_n \end{aligned} \tag{6.4}$$

where x_i is the ith analytical measurement of the concentrations c_1 to c_n and γ_{ik} are the *partial sensitivities* for each measurement. [The use of these equations implies (1) linearization by suitable transformation of the variables and (2) the mathematical independence of c and x.] The matrix

$$\begin{bmatrix} \gamma_{11} & \gamma_{12} & \cdots & \gamma_{1n} \\ \gamma_{21} & \gamma_{22} & \cdots & \gamma_{2n} \\ \gamma_{n1} & \gamma_{n2} & \cdots & \gamma_{nn} \end{bmatrix} \tag{6.5}$$

is called the *calibration matrix*. A particular analytical procedure would be termed *fully selective* when the elements of the principal diagonal are the only nonzero values. By extension, *full specificity* is defined by having only one nonzero value of γ_{ik}.

Let us now relate the statements above to chemical sensor science and its applications to multicomponent gas-phase samples. A calibration matrix for the application of a battery of chemical sensors to a multicomponent sample can be envisaged. For example, we could think of addressing a problem through use of semiconductor SnO_2, acoustic-wave, and chemiresistor devices. Accordingly, the various measurements (x_i) would incorporate a set of partial sensitivities. In addition to this form of selectivity—which is derived from the underlying physics of the device concerned—the film deposited on the sensor also confers a "chemical" form of selectivity. The latter is viewed as being entirely generated by the interaction energy for the binding site with the particular molecule concerned (Section 5.1). Accordingly, if detection of a single molecule in a complex matrix is required—such as a nitroorganic compound in the air surrounding a suitcase—it is desirable to manipulate both the device physics *and* the chemical coating in order to attempt generation of a single nonzero γ_{ik} value. On the other hand, it may be appropriate to set up an array of sensors to study the complex mixture. To date, most forays into this area have been concerned with the utilization of identical sensors (such as SnO_2 or SAW structures) where the calibration matrix has been defined entirely by the variety of binding films coated on the array of devices. This approach, and the mathematical techniques employed to achieve pattern recognition, constitutes the following section.

6.4.2. Pattern Recognition

We assume here that a sensor array produces a set of signals in response to a collection of known chemical entities (supervised learning) or to an unknown complex mixture (unsupervised configuration). Clearly, each sensor will elicit a response that is an additive function of its interaction with the various components of the mixture. These signals are then treated by a number of chemometric procedures.

The first approach to be outlined is *principal components analysis* (PCA), which is a powerful, linear supervised-learning pattern-recognition method (50). The purpose of PCA is to reduce the high dimensionality of a multivariate problem in which the variables are partly correlated to two or more dimensions by searching for qualitative associations between the variables. The association is measured by a principal component, where the procedure involves a line which best fits a system of points in space. The *first principal component* seeks to summarize as much of the joint variation in the data as possible. *The second principal component,* perpendicular to the first, seeks to summarize as much of the remaining variation, and so on. The different principal components are described by linear equations which express each component as a linear combination of the (weighted) observed variables. They are calculated as eigenvectors in such a way as to minimize the reduced error at each stage. The eigenvalue gives a measure of the amount of variance accounted for by that principal component.

Cluster analysis, which is an unsupervised technique for pattern recognition, is often employed in conjunction with PCA to identify groups of points (x_{ik}) in space (51). The analysis calculates the Euclidean distance defined by

$$d_{ij} = \sqrt{\sum_{k=1}^{N} (x_{ij} - x_{jk})^2} \tag{6.6}$$

The closeness of particular points is then calculated from the similarity index (S_{ij}):

$$S_{ij} = 1 - \frac{d_{ij}}{\max(d_{ij})} \tag{6.7}$$

Having computed the parameter, a number of techniques can then be employed to link the various points. The end result is often displayed as a dendogram, such as that illustrated in Figure 6.8.

Finally, the last technique we introduce in our brief discussion of pattern recognition methods is the *artificial neural network* (ANN) (52). (The reader might also consult reference 53 for a comprehensive and readable treatment of this rapidly growing field). A neural network consists of a number of interconnected elements or *neurons* where the strengths of the connections are adjusted (trained) to achieve an output. The latter is determined by a learning algorithm. The link between this structure and the architecture of the human brain has spawned the term neural network, although it is obvious that there are numerous differences between the artificial configuration and the biological equivalent. Historically, the field received an enormous boost by the work of Minsky and Papert on the perceptron in the 1950s and 1960s. A lull followed, but recent times have seen tremendous interest in ANN applications in engineering and science in terms of process control and scientific analysis.

In general terms, there are two classifications of network—the feedforward and recurrent configurations. These can be subdivided into specific networks, such as those invented by Hopfield, Jordan, and Kohonen. Here, we

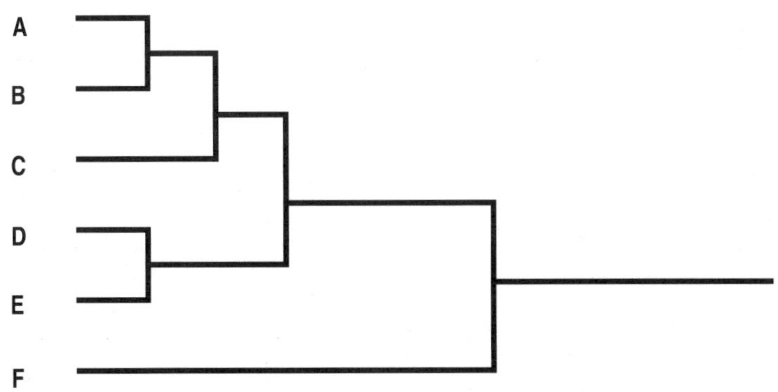

Figure 6.8. Typical dendrogram obtained by cluster analysis on a five-component system.

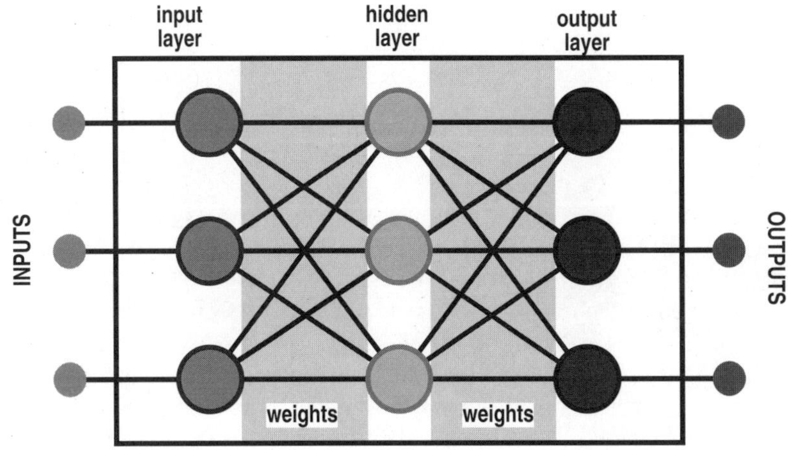

Figure 6.9. Example of a neural net featuring three nodes in a hidden layer.

show a simple three-layer feedforward network (Fig. 6.9) which incorporates *input, hidden,* and *output* layers. The structure can be operated in the *background error propagation* or *backpropagation mode.* In this procedure, input patterns presented to the input layer—say signals from an array of chemical sensors—generates a flow of activation to the output layer. Errors in the output are then fed back to the input layer to modify the weights of the interconnects. It should be emphasized that backpropagation does not describe a network, rather it represents a learning algorithm. In this way, the network can be trained with known parameters, such as sensor-array responses to sets of known chemicals. The network can then "recognize" an unknown chemical composition when the sensor array is challenged with such a cocktail.

6.4.3. Arrays of Acoustic Wave Sensors

In this section we take the liberty of including arrays involving the TSM device in addition to SAW sensors, because the amount of work done so far is minimal and, furthermore, the ANN has not yet been employed for the latter.

The first effort to apply pattern recognition techniques to sensor arrays was the work of Kowalski and co-workers in 1986 (54). In this example, 27 different gas-chromatographic stationary-phase materials were coated on different TSM sensors before exposure to 14 analytes. With respect to the responses of the sensors, PCA was used to extract the eigenvalues and eigenvectors from the covariance matrix as described briefly above. The data were also subjected to hierarchical components analysis in order to perform a comparison with the results of PCA and to examine a possible classification system for the various phases. The main goal of this work was to find a subset of films that would describe 95% of the information, and in this the authors were quite successful.

This work was followed in the same year with work on an array of 12 SAW devices exposed to 11 different chemical vapors constituting two classes, namely, polar and nonpolar (55). Here again there was an emphasis on chromatography-like stationary phases together with the solubility properties of the coatings, the overall purpose being to make the best choice of film(s) for a particular analytical sensor problem. As was the case for the work described above, the various responses were analyzed by both PCA and hierarchical CA. The former shows that members of each class of vapors tend to cluster together, which might be expected in view of their polarity. From an overall analysis of the data, it was shown that four coatings could easily separate polar and nonpolar gases, a similar result to that found by Kowalski et al. (54).

Variable solubility of compounds in a film deposited on sensors was also the subject of a device-array study by Karube et al. (56). In this case, a number of lipid species were coated onto different SAW resonator devices prior to exposure to a number of odorant molecules such as menthone and amyl acetate. The responses of the four coatings to the eight vapors employed were each normalized and then compared with a simple bar diagram. The apparent thrust of this work was to contribute to the understanding of the process of olfactory reception. The work ignores the general consensus held for some time, however, that signalling in olfaction is mediated by protein-based receptor binding (57).

Recently, Zellers et al. (58) worked with a number of polymers deposited on SAW devices and measured their responses to six organic vapors belonging to three categories. Cluster analysis and extended disjoint principal components analysis with Monte Carlo simulation were used to demonstrate that four polymer coatings could generate an 87% rate of prediction of vapor identity. Important practical issues were resolved in this study, in that responses were shown to be additive for multicomponent samples and polymer films were stable for relatively long periods of time. The crucial point was raised, however, that difficulties may ensue if the array is exposed to a compound or compounds *not* employed in the learning exercise. Similar recognition of organosulphur and organophosphorous vapor was achieved by use of a temperature-controlled array of SAW devices (59). In this particular study, sample precalibration was incorporated into the experimental design. Interestingly, the phrase "smart sensor" was involved to give something of the flavor of the configuration for identifying and quantifying various vapors.

Finally, we turn to the limited amount of research performed on the responses of acoustic-wave devices using ANN. An excellent example of the possibilities presented by the neural network is the study of whisky aromas carried out by Moriizumi and his group in recent years (60). Figure 6.10 depicts a schematic of the experimental arrangement. In this configuration an array of TSM devices with coatings of various lipid and polymer membranes in place was exposed to a number of whiskey vapors. The data was analyzed by a feedforward–backpropagation three-layer neural network which incorporated eight input, seven hidden, and five output neurons. The most interesting aspect of this work is the attempt to use the full power of the ANN in the true learning sense. No attempt was made to present the experimental array of sensors with a set of known chemical compounds, as outlined for the studies mentioned above. In this case, the network was trained with a number

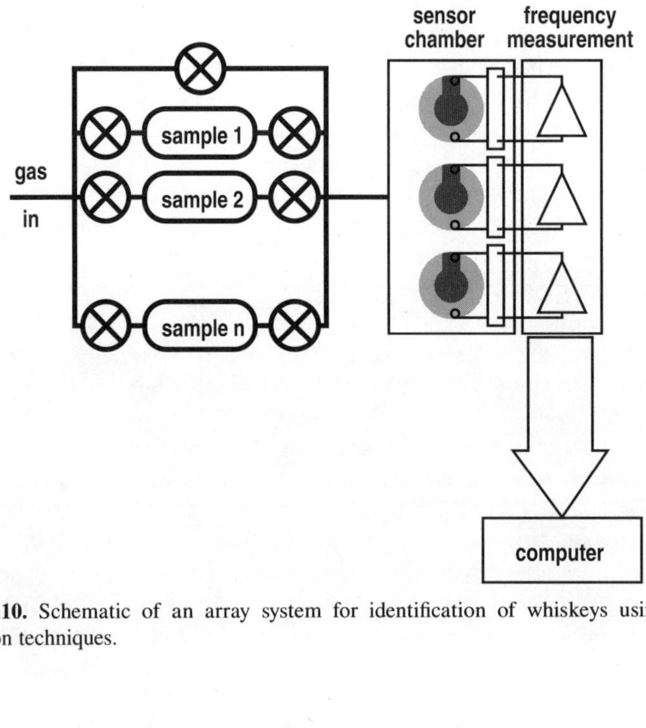

Figure 6.10. Schematic of an array system for identification of whiskeys using pattern-recognition techniques.

of known whiskeys and then employed to identify unknown alcoholic beverages.

6.4.4. Sensor Arrays as a Metaphor for the Human Olfactory System—The "Electronic Nose"

As we have seen above, arrays of various acoustic-wave devices have been employed to characterize multicomponent gas mixtures. Not surprisingly, a number of other types of sensor array, such as those based on semiconductor and chemiresistor structures, have been used with essentially the same aim. Accordingly, we address a few remarks to the specific study of odorants and aromas, where it is quite possible that commercial instruments may appear in the near future.

From a practical point of view, odors are extremely important in a number of industries, including foodstuffs and perfumes. They are often composed of quite a large number of different organic compounds with molecular weights up to approximately 300 Daltons. The interest in characterizing such odors, and the developments in sensor-array technology in conjunction with pattern-recognitive methods, is currently leading to strong commercial progress in the design and production of so-called "electronic noses." The use of the latter term is commercially attractive and tempting in view of the parallels between this technology and the human olfactory system. It is also advisable to keep in mind, however, that the biological version is radically different in many ways and is, of course, backed up by that amazing structure, the human brain. In brief, the olfactory process involves trapping of stimulants in the olfactory mucosa (compare to modern-day chromatography and solid-phase extraction) where direct interaction with receptor proteins embedded in the membrane of olfactory neurons occurs. An interesting feature of the olfactory neuron is that it not only binds a particular stimulant molecule directly, it also possess the necessary ion channel machinery to develop an action potential. These signals are then passed to highly complex arrays of other cells, in the glomerulus for example, for processing. Apparently, the stimulation of the olfactory epithelium is essentially mapped into two dimensions for processing, although we still have a lot to learn about how the overall system is able to detect and characterize complex vapors.

Readers will be able to judge for themselves whether the term "electronic nose" is justified. Perhaps what is more important than sterile arguments over terminologies, are the enormous opportunities offered by the genuine mimicking of biological phenomena. Clearly, a great deal can be learned from understanding molecular receptor binding of small molecules, receptor protein conformational changes, production of action potentials, and neural processing, among a number of possibilities.

6.5. FUTURE PERSPECTIVES

This is where we get to polish our crystal ball in an attempt to envision something of the future for the type of devices described in this text. At the risk of looking foolish in a few years, we will try to pinpoint particular areas of development, including theoretical and practical aspects of the field.

Let us begin with a few remarks about chemical and biosensor technology in general. An indicator of the periods of development for these two areas is the occurrence of the two World Conferences held in 1996 (Bangkok, Fourth World Congress on Biosensors, and Gaithersberg, Sixth Conference on Chemical Sensors). Despite the considerable effort directed at commercial development of these devices, we have yet to see their widespread application. To be sure, there have been a number of successes, an example being the surface-plasmon resonance instrument from Pharmacia, but it is fair to say that the expectations of the 1980s have yet to be fully realized. Clearly, there are technical problems in the application of chemical and biosensors in general, and acoustic-wave devices present no exception to these difficulties. This leads us to contemplate the nature of the areas that have held up the general application of sensors, and what might be done to address these problems in the future.

One of the very attractive features of sensor technology is the potential ability to make continuous measurements over time of concentrations of analytes, examples being ion concentration in blood and levels of noxious gases in the workplace. But this, to some degree, is where the "Achilles Heel" of the technology lies. It has proven to be a difficult task to achieve continuous calibration of sensor response and, accordingly, to relate precise concentrations in the gas or liquid phase to the various types of electrical response generated by devices. The enormous and costly effort of developing a closed-loop detector configuration for glucose in the blood of diabetic patients constitutes one extreme example in this area. So what is the nature of the problem? In our view, the long-term stability of the surface chemistry imposed on a particular device has not really been addressed, which is especially critical when it comes to biological macromolecules deposited on substrates. In other words, we can never really be sure as to the quality of the concentration–signal relationship over time *if* the sensor interface to the sample matrix is changing owing to intrinsic alteration in receptor chemistry and/or fouling of the surface.

Perhaps solutions to these problems lie in the design of much more robust chemical systems and, perhaps, "self-calibration" of devices through application of chemometric methods. Although engineers would undoubtedly argue differently, it appears that devices themselves do not constitute a major difficulty in that reasonable stability can be achieved in this day and age. One consequence of the problems of surface chemistry mentioned above has been the success achieved with one-shot structures that can determine a concentration level in one experiment only. Diabetics will be all too familiar with the pricking of their anatomy with a needle prior to use of a glucose test strip. Additionally, there has been some success achieved in fields where long-term measurements are not required such as nucleic-acid hybridization and urea-blood concentrations during dialysis of patients with renal failure.

We now turn to some more specific remarks concerning acoustic-wave devices. One area in which we will certainly see considerable attention focussed is the understanding of the interaction between acoustic energy and the liquid–solid interface. It is apparent that controversy surrounds this area of physics. The debate has centered around the behavior of the TSM device in liquids, rather than the structures encountered in our text, but similar arguments could be made about the other sensors. There are two levels of discussion. First, a number of members of the acoustic-wave chemical sensor fraternity appear to want to hang on to Sauerbrey's description of changes in frequency (often not the true series resonant frequency) caused by additional mass at all cost. Despite this, in the future perhaps there will be general acceptance that the extended thickness model of Sauerbrey does not apply when considerable acoustic energy is being transferred to the liquid from the device. This is especially true when the added or subtracted mass at the device surface involves biological macromolecules. Second, the role played by the interface in the propagation of acoustic energy is likely to be much better understood. This is particularly important for devices that involve transverse waves such as the TSM, SH-APM, and STW sensors. Here, the signal generated by the device when immersed in liquids is governed by a complex array of parameters such as rugosity, viscosity, surface change, and free energy. The controversy here is concerned with the possible role played by two-dimensional viscosity or molecular slip. Arguments for and against have appeared in the literature. Resolution of this debate in the coming months will clearly lead to exciting new applications for transverse-wave devices in the study of interfacial physical chemistry.

A related area is the relationships that hold for parameters derived from acoustic network analysis in terms of the chemistry being conducted at the liquid–device interface. As we have seen earlier, the acoustic-wave device can be treated as one component of an electrical system and, therefore, an equivalent circuit can be calculated, leading to parameters such as motional resistance, inductance, and capacitance. The future is likely to see much greater effort to adduce the meaningfulness of these terms to surface chemistry. Again, such an increased understanding will lead to new applications for the examination of interfacial chemistry.

One topic left virtually untouched is that of hybrid systems and true hybrid sensors. The former refers to a configuration of sensors based on different physics being employed in concert with each other, whereas the latter involves the same device being interrogated by different physical techniques. It is possible that some of the problems alluded to above could be addressed by both methods. To give an example of a hybrid system, we describe briefly the combination of devices based on measurements of conductivity with acoustic-wave sensors. Various films based on conducting polymers such as polypyrrole can be imposed between the electrodes of an interdigital electrode array on a chemiresistor *and* on the surface of a TSM or SAW device. Exposure of both types of sensor to the same gas, for example, could lead to an enhanced picture for analysis by the chemometric methods outlined above. Such an approach may be much more feasible than one involving a single device.

The hybrid sensor may also be important from the standpoint of understanding interfacial effects. Some possibilities are the combined use of optical and pyroelectric techniques with acoustic-wave structures. For example, it is well-known that certain materials, such as lithium niobate, exhibit both piezoelectric and pyroelectric effects. Accordingly, it should be possible to multiplex these processes for examination of the chemistry conducted on a single device. There may be advantages to be gained here with respect to correction for nonspecific adsorption and other undesirable effects.

Other examples of hybrid technologies with potential sensor applications include the use of laser interferometric measurement of acoustic-wave velocities and the combination of optical (e.g., FTIR-ATR) and acoustic-wave measurements on the same physical substrate. The former technique (currently used for noninvasive evaluation of acoustic properties and defect location) may be combined with surfaces having graduated properties or multiple coatings to, in effect, establish a sensor array on a single substrate and with a single pair of IDTs. Thus array measurements could be achieved by rapid switching of the area interrogated by the interferometer, rather than by switching between physical devices or circuits.

Finally, one area that will almost certainly enjoy rapid progress is that of the characterization of the properties of thin films. Throughout this text we have seen a number of cases where the deposition of a thin film on an acoustic wave sensor has resulted in an understanding of properties of the film itself through direct study of sensor response. Such a method is likely to prove popular, for example, in the determination of viscoelastic properties as they relate to thin layers, for example, paint, polymer coatings, sol-gel chemistry, and biological macromolecules, among a number of possibilities. Progress in the study of thin films is likely to outstrip developments in the chemical sensor field. In this respect, we are greatly encouraged by the choice of the Royal Society of Chemists, UK, to hold a Faraday Division Discussion on the Interaction of Acoustic Waves with Thin Films and Interfaces, scheduled for 1997.

REFERENCES

1. *Sensors and Sensory Systems for an Electronic Nose,* J. W. Gardner and P. N. Bartlett, Eds., *NATO ASI Series, Series E; Applied Sciences,* Vol. 212, Kluwer Academic Publishers, Dordrecht, The Netherlands, 1992.
2. A. J. Ricco, S. J. Martin, and T. E. Zipperian, *Sens. Actuators,* **1985,** *8,* 319.
3. A. W. Snow, W. R. Barger, M. Klusty, H. Wohltjen, and N. L. Jarvis, *Langmuir,* **1986,** *2,* 513.
4. M. S. Nieuwenhuizen and A. W. Barendsz, *Electron. Lett.,* **1986,** *22,* 184.
5. R. Lec, J. F. Vetelino, R. S. Falconer, and Z. Xu, *IEEE Ultrason. Symp. (Proc.),* **1988,** 585.
6. M. S. Nieuwenhuizen, A. J. Nederlof, and A. W. Barendsz, *Anal. Chem.,* **1988,** *60,* 230.
7. B. Holcroft and G. G. Roberts, *Thin Solid Films,* **1988,** *160,* 445.
8. M. Rapp, I. Kabbe, M. Von Schicklus, S. Hunklinger, H. Fuchs, W. Schrepp, and B. Fleischmann, *Sens. Actuators B,* **1991,** *B4,* 103.
9. W. J. Ghijsen, A. W. Barendsz, and M. S. Nieuwenhuizen, *IEEE Trans. Ultrason. Ferroelec. Freq. Control,* **1987,** *UFFC-34,* 149.
10. A. D'Amico, A. Palma, and E. Verona, *IEEE Ultrason. Symp. (Proc.),* **1982,** 308.
11. A. D'Amico, A. Palma, and E. Verona, *Appl. Phys. Lett.,* **1982,** *41,* 300.
12. A. Bryant, M. Poirier, G. Riley, D. L. Lee, and J. F. Vetelino, *Sens. Actuators,* **1983,** *4,* 105.
13. R. L. Cook, R. C. MacDuff, and A. F. Sammells, *Anal. Chim. Acta,* **1989,** 217, 101.
14. J. F. Vetelino, R. K. Lade, and R. S. Falconer, *IEEE Trans. Ultrason. Ferroelec. Freq. Control,* **1987,** *UFFC-34,* 56.
15. A. Snow and H. Wohltjen, *Anal. Chem.,* **1984,** *56,* 1411.
16. T. Bein, K. Brown, G. C. Frye, and C. J. Brinker, *J. Am. Chem. Soc.,* **1989,** *111,* 7640.

17. F. L. Dickert, A. Haunschild, M. Reif, and W. E. Bulst, *Adv. Mater.*, **1993,** *5,* 277.
18. E. T. Zellers, N. Hassold, R. M. White, and S. M. Rappaport, *Anal. Chem.*, **1990,** *62,* 1227.
19. E. T. Zellers, R. M. White, and S. M. Rappaport, *Anal. Chem.*, **1990,** *62,* 1222.
20. E. T. Zellers and G.-Z. Zhang, *Anal. Chem.*, **1992,** *64,* 1277.
21. G.-Z. Zhang and E. T. Zellers, *Anal. Chem.*, **1993,** *65,* 1340.
22. E. T. Zellers and G.-Z. Zhang, *Anal. Chim. Acta,* **1993,** *280,* 1.
23. G. Watson and E. Staples, *IEEE Ultrason. Symp. (Proc.),* **1990,** 311.
24. G. Watson, W. Horton, and E. Staples, *First Intl. Symp. Explosive Det. Tech. (Proc.),* **1991,** 589.
25. H. Wohltjen and R. E. Dessy, *Anal. Chem.*, **1979,** *51,* 1458.
26. H. Wohltjen and R. E. Dessy, *Anal. Chem.*, **1979,** *51,* 1465.
27. H. Wohltjen and R. E. Dessy, *Anal. Chem.*, **1979,** *51,* 1470.
28. H. Muramatsu, E. Tamiya, M. Suzuki, and I. Karube, *Anal. Chim. Acta,* **1989,** *217,* 321.
29. B. J. Costello, S. W. Wenzel, A. W. Wang, and R. M. White, *IEEE Ultrason. Symp. (Proc.),* **1990,** 279
30. A. W. Wang, B. J. Costello, and R. M. White, *Anal. Chem.*, **1993,** *65,* 1639.
31. S. J. Martin and A. J. Ricco, *Sens. Actuators A*, **1990,** *A21-A23,* 712.
32. R. C. Hughes, S. J. Martin, G. C. Frye, and A. J. Ricco, *Sens. Actuators A* , **1990,** *A21-A23,* 693.
33. J. W. Grate, S. W. Wenzel, and R. M. White, *Anal. Chem.*, **1992,** *64,* 413.
34. S. L. Hietala, P. M. Smith, V. M. Hietala, G. C. Frye, and S. J. Martin, *Langmuir,* **1993,** *9,* 249.
35. D. Li and B. I. Swanson, *Langmuir,* **1993,** *9,* 3341.
36. J. E. Roederer and G. J. Bastiaans, *Anal. Chem.*, **1983,** *55,* 2333.
37. G. S. Calabrese, H. Wohltjen, and M. K. Roy, *Anal. Chem.*, **1987,** *59,* 833.
38. M. Thompson, C. L. Arthur, and G. K. Dhaliwal, *Anal. Chem.*, **1986,** *58,* 1206.
39. M. Thompson, G. K. Dhaliwal, C. L. Arthur, and G. S. Calabrese, *IEEE Trans. Ultrason. Ferroelec. Freq. Control,* **1987,** *UFFC-34,* 128.
40. Biosensors '96, Fourth World Cong. Biosensors, (Abstract) Elsevier Advanced Technology, Amsterdam, 1996.
41. B. J. Costello, B. A. Martin, and R. M. White, *IEEE Ultrason. Symp (Proc.),* **1989,** 977.
42. J. Andle, J. Vetelino, M. Lade, and D. McAllister, *IEEE Ultrason. Symp. (Proc.),* **1990,** 291.
43. E. Gizeli, N. J. Goddard, C. R. Lowe, and A. C. Stevenson, *Sens. Actuators B,* **1992,** *B6,* 131.
44. G. Kovacs, M. J. Vellekoop, R. Haueis, G. W. Lubking, and A. Venema, *Sens. Actuators A.,* **1994,** *A43,* 38.

45. J. Kondoh, Y. Matusi, S. Shiokawa, and W. B. Wlodarski, *Sens. Actuators B,* **1994,** *B20,* 199.

46. M. Tom-Moy, R. L. Baer, D. Spira-Solomon, and T. P. Doherty, *Anal. Chem.,* **1995,** *67,* 1510.

47. M. Tom-Moy, T. P. Doherty, R. L. Baer, and D. Spira-Solomon, in *Biosensor and Chemical Sensor Technology,* K. R. Rogers, A. Mulchandani, and W. Zhou, Eds., ACS Symposium Series, Vol. 613, American Chemical Society, Washington DC, 1995, p. 9.

48. J. Renken, R. Dahint, M. Grunze, and F. Josse, *Anal. Chem.,* **1996,** *68,* 176.

49. H. Kaiser, *Pure. Appl. Chem.,* **1973,** *34,* 35.

50. E. R. Malinkowski and D. G. Howery, *Factor Analysis in Chemistry,* Wiley, New York, 1980.

51. D. L. Massart and L. Kaufman, *The Interpretation of Analytical Chemical Data by the Use of Cluster Analysis,* Wiley, New York, 1983.

52. P. A. Jannson, *Anal. Chem.,* **1991,** *63,* 357A.

53. M. Caudill and C. Butler, *Naturally Intelligent Systems,* MIT Press, Cambridge, MA, 1990.

54, W. P. Carey, K. R. Beebe, B. R. Kowalski, D. L. Illman, and T. Hirschfeld, *Anal. Chem.,* **1986,** *58,* 149.

55. D. S. Ballantine, S. L. Rose, J. W. Grate, and H. Wohltjen, *Anal. Chem.,* **1986,** *58,* 3058.

56. S. M. Chang, B. Ebert, E. Tamiya, and I. Karube, *J. Biotech.,* **1990,** *16,* 211.

57. *Chemosensory Information Processing,* D. Schild, Ed., Springer, Berlin, 1990.

58. E. T. Zellers, S. A. Batterman, M. Han, and S. J. Patrash, *Anal. Chem.,* **1995,** *67,* 1092.

59. J. W. Grate, S. L. Rose-Pehrsson, D. L. Venezky, M. Klusty, and H. Wolhltjen, *Anal. Chem.,* **1993,** *65,* 1868.

60. Reference 1, T. Moriizumi, T. Nakamoto, and Y. Sakuraba, p. 217.

APPENDIX

A

GLOSSARY OF SYMBOLS

a Finger separation

A Area

b Unit cell dimension of an IDT ($= d + a$)

B_a Radiation susceptance (acoustic origin)

C Capacitance

C_1 Proportionality term in the Sauerbrey equation

C_m Mass sensitivity of an acoustic wave device

C_s Static capacitance (per unit length)

C_T Transducer capacitance

C_v Vapor-phase concentration

$\mathbf{c}$ Elastic coefficient tensor

$\boldsymbol{D}$ Electric displacement vector

d IDT finger width

$\mathbf{d}$ Strain-coefficient tensor

$\mathbf{E}$ Electric field vector

$\mathbf{e}$ Piezoelectric stress-coefficient tensor

$\boldsymbol{F}$ Force

F Difference between sample and reference frequencies

ΔF Change in the difference frequency

f Frequency

f_0 Unperturbed resonance frequency

Δf Change in frequency

G Complex shear modulus

G_a Electrical radiation conductance (acoustic origin)

h Film thickness

K^2 Coupling coefficient

k Wavenumber($= 2\pi/\lambda$)

K Complex bulk modulus

L Delay line length (center-to-center IDT separation); *also* inductance

l Length or separation

m Mass

m_f Mass of film or deposit

Δm Change in mass

M Mass per unit area

ΔM Change in mass per unit area

N Number of finger pairs (IDT)

N_v Number of molecules per unit volume

N_λ Number of acoustic wavelengths in the propagation path

p Partial pressure of a vapor

p_0 Saturated vapor pressure

P Pressure

$\boldsymbol{P}$ Polarization vector

q Volumetric flow rate

r Metallization ratio of an IDT ($= d/b$)

R Resistance

$\boldsymbol{S}$ Strain vector

S_m Mass sensitivity

S_m^V Mass sensitivity measured at constant frequency

s Compliance coefficient tensor

t Thickness

T Temperature

$\boldsymbol{T}$ Stress vector

u Particle displacement

v Instantaneous voltage

V Acoustic-phase velocity

V_0 Unperturbed acoustic-phase velocity

ΔV Change in acoustic-phase velocity

w Width

W Beam width (IDT)

Y Electrical admittance

z Arbitrary distance

Z Electrical or acoustic impedance

α Attenuation (imaginary part of the complex wavenumber)

β Real part of the complex wavenumber

δ Number of molecules per unit area; *also* evanescent decay length

$\boldsymbol{\varepsilon}$ Dielectric strain coefficient tensor

ε_0 Permittivity of free space

ε_r Relative permittivity of a dielectric medium

ε_p^T Stress-free relative permittivity

η Viscoelastic coefficient tensor

κ Fraction of area that is acoustically active

λ Acoustic wavelength

λ' First Lamé constant

μ_D Dipole moment

μ_s Substrate shear modulus

μ' Second Lamé constant

ρ_f Fluid density

ρ_m Material density

ρ_s Substrate density

ρ_{sc} Surface charge density

σ Surface (sheet) conductivity

υ Wave propagation velocity

ϕ Electric potential wave

Φ_E Electrical circuit-phase shift

χ Polarizability (or electronic susceptability)

ω Angluar frequency ($= 2\pi f$)

INDEX